THE HISTORY OF
SCIENCE AND TECHNOLOGY
IN THE UNITED STATES

BIBLIOGRAPHIES OF THE HISTORY
OF SCIENCE AND TECHNOLOGY
(Vol. 2)

GARLAND REFERENCE LIBRARY
OF THE HUMANITIES
(Vol. 308)

Volume 2

Bibliographies of the
History of Science and Technology

Editors

Robert Multhauf
Smithsonian Institution, Washington, D.C.

Ellen Wells
Smithsonian Institution, Washington, D.C.

THE HISTORY OF
SCIENCE AND TECHNOLOGY
IN THE UNITED STATES
A Critical and Selective Bibliography

Marc Rothenberg

GARLAND PUBLISHING, INC. • NEW YORK & LONDON
1982

Library of Congress Cataloging in Publication Data

Rothenberg, Marc, 1949–
   The history of science and technology in the United States.

   (Bibliographies of the history of science and technology ; v. 2) (Garland reference library of the humanities ; v. 308)
   Includes indexes.
   1. Science—United States—History—Bibliography.
2. Technology—United States—History—Bibliography.
I. Title.   II. Series.   III. Series: Garland reference library of the humanities ; v. 308.
Z7405.H6R67   1982   [Q125]   016.50973   81-43355
ISBN 0-8240-9278-3

Printed on acid-free, 250-year-life paper
Manufactured in the United States of America

GENERAL INTRODUCTION

This bibliography is one of a series designed to guide the reader into the history of science and technology. Anyone interested in any of the components of this vast subject area is part of our intended audience, not only the student, but also the scientist interested in the history of his own field (or faced with the necessity of writing an "historical introduction") and the historian, amateur or professional. The latter will not find the bibliographies "exhaustive," although in some fields he may find them the only existing bibliographies. He will in any case not find one of those endless lists in which the important is lumped with the trivial, but rather a "critical" bibliography, largely annotated, and indexed to lead the reader quickly to the most important (or only existing) literature.

Inasmuch as everyone treasures bibliographies, it is surprising how few there are in this field. Justly treasured are George Sarton's *Guide to the History of Science* (Waltham, Mass., 1952; 316 pp.), Eugene S. Ferguson's *Bibliography of the History of Technology* (Cambridge, Mass., 1968; 347 pp.), François Russo's *Histoire des Sciences et des Techniques. Bibliographie* (Paris, 2nd ed., 1969; 214 pp.), and Magda Withrow's *ISIS Cumulative Bibliography. A bibliography of the history of science* (London, 1971–; 2131 pp. as of 1976). But all are limited, even the latter, by the virtual impossibility of doing justice to any particular field in a bibliography of limited size and almost unlimited subject matter.

For various reasons, mostly bad, the average scholar prefers adding to the literature, rather than sorting it out. The editors are indebted to the scholars represented in this series for their willingness to expend the time and effort required to pursue the latter objective. Our aim has been to establish a general framework which will give some uniformity to the series, but otherwise to leave the format and contents to the author/compiler. We have urged that introductions be used for essays on "the state of the field," and that selectivity be exercised to limit the length of each volume to the economically practical.

Since the historical literature ranges from very large (e.g., medicine) to very small (chemical technology), some bibliographies will be limited to the most important writings while others will include modest "contributions" and even primary sources. The problem is to give useful guidance into a particular field—or subfield—and its solution is largely left to the author/compiler.

In general, topical volumes (e.g., chemistry) will deal with the subject since about 1700, leaving earlier literature to area or chronological volumes (e.g., medieval science); but here, too, the volumes will vary according to the judgment of the author. The topics are international, with a few exceptions (Greece and Rome, Eastern Asia, the United States), but the literature covered depends, of course, on the linguistic equipment of the author and his access to "exotic" literatures.

Robert Multhauf
Ellen Wells

Smithsonian Institution
Washington, D.C.

CONTENTS

PREFACE

This volume is a guide to the secondary literature in the history of American science and technology, excluding the history of medicine. It has been compiled to assist newcomers—whether graduate students or experienced scholars from other segments of the history of science and technology or American history—in entering an exciting and growing field of historical research. This is not meant to be a complete bibliography. Rather, it is an orientation to the literature, a general guidebook which will provide an overview, point out the most important attractions, and warn of hidden pitfalls and dangers.

With very few exceptions, the entries appeared between 1940 and 1980. Most pre-1940 imprints have been superseded. Time restraints led to the decision not to include any work which appeared post-1980. Included among the entries are books, articles, dissertations, and a review essay. All the entries are in the English language. This was not the result of a far-sighted plan but a reflection of the reality of the situation; almost all investigators of the history of American science and technology are from English-speaking nations. A few articles included in this volume were written by Continental scholars, but they were published in English. The one major monograph by a Continental writer is available in translation.

This bibliography is a distillation of over sixteen hundred works. To be included, an entry had to meet one of four criteria. First, all works considered fundamental for an understanding of the history of American science and technology, either because of content or historiographic thrust, are present. These are the books and articles any scholar with a pretension of knowledge of this field must know. Almost all the older imprints come under this criterion. In some cases these classics are now viewed as misleading or inaccurate; in other cases the historiography is no longer in fashion. In all cases, however, these books and articles are the foundation stones upon which all other historical activities have been built.

Included under the second criterion are the best works. An attempt was made to identify the most significant themes and select the most accurate, inclusive works on those themes. This effort to identify the best works was also used to select biographies for inclusion. A number of significant figures in American science and technology are represented in the literature by two or more biographies of considerable variation in quality. An arbitrary rule was invoked for dissertations. They were included only if they were not represented, in whole or in part, in the published literature.

In order to make this bibliography as comprehensive as possible within the size limitations, the third criterion used was variety. Entries were selected to demonstrate the great range of methods and sources used by historians of American science and technology and the great range of problems they investigate. Included are social history and intellectual history, studies of institutions and studies of experiments, studies of technological hardware and analyses of the social impact of that hardware.

The last criterion is one which, unfortunately, few compilers of bibliographies use. To save the neophyte from wasting his or her time reading material which is inaccurate, undocumented, or of questionable validity, I have included some entries in order to wave the red danger flag. These works have seductive titles and, in some instances, seductive themes, but they are examples of bad scholarship. I also raised the red flag when titles were misleading or inaccurate.

Excluded from this bibliography were works in the sociology of science and science policy studies. Deciding whether a particular book or article should be classified as history of science, science policy, or sociology was sometimes very difficult. Many science policy monographs, for example, utilize a historical approach or historical evidence. In these cases I have attempted to judge where the loyalties of the author lay and the intended audience. Very often the jargon of the author was used to decide whether a work was part of the historical literature or belonged in social science.

Entries have been categorized topically. Within categories, the entries are listed alphabetically by author. Many entries could logically be placed in more than one category. No attempt has been made to ensure that each entry appears under every conceivable category. Rather, with few exceptions, I have placed the entry under what I judged to be the single most useful category, relying upon the subject index to guide the user to all entries of interest. For an indication of the possible problems the alternative approach could lead to, the user is invited to examine item 426 and its two subsequent repetitions under the categories of Mathematics and Physics.

Annotations are descriptive and frequently evaluative. Generally, annotation length reflects the complexity of the entry and the need to summarize it accurately. There is no simple correlation between the quality of a work and the length of its annotation.

As I compiled this bibliography, I incurred numerous debts. My thanks must go to Jack Marquardt and the staff of the Smithsonian Institution Libraries. Their interlibrary loan arrangements were fundamental to the success of this endeavor. I also want to express my deep appreciation to Joan F. Steiner, who tried valiantly to improve my use of the English language. Any remaining awkwardness in the wording of the annotations must be blamed on my refusal to take good advice. The staff of the Joseph Henry Papers nominated entries for the bibliography which I might have otherwise overlooked. For that I am grateful. Ellen Wells was the ideal editor—prodding but understanding of a compiler who always needed a little more time to complete the task. Finally, I would like to thank my parents, to whom I dedicate this effort with love and appreciation.

INTRODUCTION

The neophyte historian of either American science or American technology is in an enviable position. These fields contain numerous interesting questions, themes, and research topics, but they are still relatively uncrowded. Previously compiled needs lists, even those up to a quarter-century old (e.g., item 15) are filled with many suggestions for important research topics which have never generated a response. Add the subjects for which no research exists but works by historians of earlier generations, works no longer accurate or reflective of current historiography, and the topics created by the history of recent science and technology, and the sense of needs and opportunities is almost overwhelming.

Although there are many interfaces between the history of American science and the history of American technology, the two disciplines should be considered separately. The evolution of the disciplines differed sharply. So did the training of the historians. Although some historians attempt to straddle both fields, most are comfortable only in one or the other. At best they dabble in the opposite field. The differences can be seen clearly in the secondary literature.

I

Historians of American science have been blessed. Until relatively recently, there have been few outstanding American scientists. As a result, historians of American science have not felt obligated to restrict themselves to "Great Scientist" history. Instead of the "Newton industry" or the "Darwin industry" manifestations of a history of science which all too often concentrates only on a few giant trees in the forest, the American scene is dominated by networks of social and intellectual historians who perceive science as part of their purview. These Americanists have looked at the entire forest of scientists from an ecological perspective—seedlings, young trees, maturing trees, predators, and even the fertilizer. They have explored how science functions in a particular social setting, how it is funded, how it generates popular support. The paths disciplines take as they rise and develop have been carefully followed. By the standards of historians of European science, Americanists study relatively little scientific thought, concentrating instead on patronage, professionalization, and the rise of new disciplines.

Of course, not all historians of American science have been comfortable with the lack of "Great Scientists." Some have responded by denouncing American science for its deficiency. This denunciation has traditionally taken the form of an attack on nineteenth-century American science as backward, prefaced by a paean to the glories of colonial science. Benjamin Franklin is presented as a clear sign that science had reached a high level in pre-1776 America. The subsequent lack of "Great Scientists" during the nineteenth century, as compared to the leading European countries, demonstrated, according to these historians, that the relationship between American science and society was disfunctional.

A massive inferiority complex is evident in these attacks. Rather than accepting that the United States was a developing nation compared to the major European scientific powers (and that Franklin was a freak occurrence), these historians had to find a pathology—the indifference of American society to basic research. Lacking comparative data and, on occasion, clarity in their terminology, the major result of the efforts of the indifference school was the distortion of the history of American science. They produced prescriptive, not analytical or descriptive, history.

Fortunately, such an attitude is now held by only a very few. Even one of the leading advocates of the indifference argument, with its corollary of the decline of American science during the nineteenth century, has reexamined and softened his position (see items 26 and 27).

Although there were relatively few great American scientists, at least pre-1930, there have been many significant individuals. These individuals have contributed to the institutional development of science and provided important, albeit unspectacular, advances in scientific knowledge. Yet there are few scholarly biographies of these individuals. Indeed, biography seems to be a genre which has been neglected by Americanists over the last two decades. Perhaps feeling that biographies can only be written about "Great Scientists," the Americanists have carefully considered the institutional frameworks surrounding the scientist, evaluated the experiments and theories generated within these frameworks, but have been reluctant to examine the scientists themselves. For most of the American scientific community, the biographical literature is either nonexistent or limited to the hagiologic, inaccurate, or simply dated. This bibliography could have contained at least twenty-five additional monograph-length biographies or collections of biographical sketches; but these works were neither bad enough to generate a warning nor of sufficient quality to be included. For the most part, these biographies are forty or more years old. Since their publication, more manuscripts have been discovered, better interpretations have arisen, and standards have been raised.

These older biographies point to a more general problem in the history of American science—the aging of the literature. Few of the older works survive the test of time, except as historiographic documents. Pre-1940 imprints are frequently beyond their useful age, while many pre-1955 imprints are marginal. This may simply be a manifestation of each generation desiring to rewrite history. However, it is more likely a reflection of the situation in the field in the 1940s and 1950s. The contemporary reviews of books published during those decades present an interesting pattern. So little was known about the history of American science that almost any new information was welcome. Standards were, by the measure of today, relatively low. First steps, however, are always the most difficult.

There are other topics in the history of American science which have been neglected. I will focus on only two: scientific instrumentation and the social sciences.

Despite statements that instrumentation has been vital in the development of certain disciplines (e.g., item 552), Americanists have done relatively little research on scientific instrumentation and its impact on science. There are exceptions. Historians of astronomy recognize that the quality and quantity of telescopes

and auxiliary equipment govern the very questions the astronomer can ask. Microscopes have been studied with an eye for their
influence on the biomedical sciences. But there is room for con-
siderably more work in these areas, let alone the subjects of
chemical and physical apparatus. In addition, the studies of apparatus are skewed towards the colonial period and the nineteenth
century. The growth of Big Science in the twentieth century has
also been dependent in no small part on the growth of instrumentation, but the Americanists have failed to investigate that
dependency.

The social sciences present a set of unusual circumstances to
the Americanist. Of all the fields, the history of the social sciences
is the home of the most practitioner history; i.e., history written by
the scientist, frequently for consumption by the scientist. This has
resulted in history of uneven quality, Whiggish tendencies, and an
isolation from other Americanists.

Of course, it is inaccurate to view the social sciences as mono-
lithic. The history of anthropology is very much in the mainstream of the history of American science. Studies of the American School of Anthropology (the label given to the antebellum
group of physical anthropologists who argued that Blacks were
inferior to Whites), for example, represent history of science at a
very high level, as do studies of the professionalization of anthropology at the turn of the twentieth century. Sociology, in
contrast, with the exception of some of the studies of the Chicago
department, has fewer ties to mainstream history and is of lower
overall quality. The behavioral sciences are very much practitioner fields, with all of the negative manifestations.

II

The history of American technology is presently the meeting
ground of the social historian, the economic historian, and the
historian of technological hardware. None of these three groups
has a monopoly on asking interesting questions or doing history
of high quality. However, all three ask different questions.

Social historians of technology ask many of the questions
about engineers, mechanics, and technology that their counterparts in the history of science ask about the members of the

scientific community. Themes popular with social historians are the professionalization of engineers, the organization of engineering or technical research and development, patronage, and the interaction of the technologist with the wider society. Economic historians have focused on the business organization and its utilization of innovation. The interaction of organizational evolution, changes in marketing conditions, and technical innovation fascinate these historians.

The Americanists who write about hardware account for the development of such artifacts as bridges, automobiles, airplanes, and machine tools. As done by the best of these historians, the result is not a recitation of technical changes but a deep understanding of how inventors respond to technical, social, and economic problems. Unfortunately, few works meet such standards.

The interests of the three groups coalesce in the history of technological change. In what is the most difficult form of the history of technology, the scholar must synthesize the knowledge generated by the tools of social, economic, and technical history to explain the fundamental problem in the history of technology. Technological change involves invention, development, diffusion, and transfer. Engineers, entrepreneurs, scientists, and government officials are often part of the process.

One very special form of technological change which has preoccupied historians of American technology has been the origin and development of the American system of manufacturing—the mass production of interchangeable parts. The American system substituted resources for labor, sacrificed quality for quantity, and delivered cheap goods in mass quantities. It helped establish the uniqueness of the material aspect of the American way of life, and no detail of it appears insignificant to the Americanist.

Unlike the history of American science, the history of American technology has been inflicted with great men—in this case, the "Heroic Inventor." However, Americanists have quickly come to terms with the "Heroic Inventor." In many cases they have demonstrated that he was more myth than reality. In other cases they have built up a structure around him, presenting this individual as the head of an organization with a variety of support facilities, not a solitary figure against the world. Americanists often follow-up their deflation of the "Heroic Inventor" by ignoring him in favor of more significant problems.

Historians of American technology do share a dilemma with historians of American science: what to do with the older, hagiographic biographies. Their response to this problem has been somewhat more forceful. A new generation of biographies is appearing to take the place of the older works.

At the same time, the problem of aging literature does not appear to be as great in the history of technology as in its counterpart. At least, the products of the economic historians created during the 1940s and 1950s have withstood the test of time rather well. The explanation must lie, at least in part, in the fact that economic history was better established and could demand higher standards than the history of American science.

In other ways, Americanists interested in technology have a neater field than their counterparts. There is no disaster area akin to the history of the social sciences in the history of technology, although the history of the National Aeronautics and Space Administration, at least in its early manifestations, comes very close. While there are needs to be filled, no obviously important but totally neglected area jumps out to catch the eye.

III

As a postscript to earlier remarks about the growth and opportunities in the history of American science and technology, I want to point out that during the first ten months of 1981 some fifty articles, monographs, and dissertations appeared which could have been included in this bibliography. For reasonably up-to-date information on these fields, the reader is urged to monitor the annual issues of items 5 and 6.

THE HISTORY OF
SCIENCE AND TECHNOLOGY
IN THE UNITED STATES

CHAPTER I: BIBLIOGRAPHIES AND GENERAL STUDIES

BIBLIOGRAPHIES

1. Black, George W., Jr. *American Science and Technology: A Bicentennial Bibliography*. Carbondale and Edwardsville: Southern Illinois University Press, 1979. xi + 170 pp. Indices.

    Lists articles published in 1976. Although many entries are not historical in nature, this work is useful in locating narrow-focus articles.

2. Caldwell, Lynton K. *Science, Technology and Public Policy: A Selected and Annotated Bibliography*. 3 Volumes. Bloomington: Program in Public Policy for Science and Technology, Department of Government, Indiana University, 1969-1972. xii + 492 pp., xii + 544 pp., xii + 868 pp. Indices.

    Limits coverage to English language publications published after 1944. Provides author and title indices, content and analytic annotations. Breaks down entries by geographical units.

3. Cutcliffe, Stephen H., Judith A. Mistichelli, and Christine M. Roysdon. *Technology and Values in American Civilization: A Guide to Information Sources*. American Studies Information Guide Series, Volume 9. Detroit: Gale Research Company, 1980. xviii + 704 pp. Indices.

    Includes over twenty-four hundred entries, almost all with descriptive, evaluative annotations. The only major deficiency of this bibliography is the failure to include items on science-technology interaction.

4. Ferguson, Eugene S. *Bibliography of the History of Technology*. Cambridge, Mass. and London: The M.I.T. Press, 1965. xx + 347 pp. Index.

    Provides both critical and descriptive annotations. This is an essential reference tool for any serious scholar in the history of technology.

5. Goodwin, Jack. "Current Bibliography in the History of Technology." *Technology and Culture*. (1964--)

    Includes chronological and subject classifications. Lacks an entry for the United States in the subject index. This bibliography has appeared annually in this journal since 1964.

6. Neu, John. "Critical Bibliography." *Isis*.

    Includes a name and author index. It is divided according to disciplines and time periods. Appears annually, with cumulative indices available (see items 7 and 8).

7. ————. *Isis Cumulative Bibliography, 1966-1975. A Bibliography of the History of Science, Formed from Isis Critical Bibliographies 91-100 Indexing Literature from 1965 to 1974.* Volume I: *Personalities and Institutions*. London: Mansell, 1980. xxix + 483 pp.

    Represents a continuation of item 8.

8. Whitrow, Magda. *Isis Cumulative Bibliography. A Bibliography of the History of Science Formed from Isis Critical Bibliographies 1-90, 1913-1965.* Volume I: *Personalities, A-J.* Volume II: *Personalities, K-Z and Institutions.* Volume III: *Subjects.* London: Mansell, 1971-1976. vii + 664 pp., 789 pp., xciv + 678 pp. Index.

    Provides an awkward classification scheme for the subject index. The coverage is uneven and incomplete for American themes. There will be a fourth volume, classifying entries according to chronological period

GENERAL SCIENCE

9. Bates, George E., Jr. "Seventeenth- and Eighteenth-Century American Science: A Different Perspective." *Eighteenth Century Studies*, 9 (1975-76): 178-192.

Contends that the premodern framework of reality was a dualistic synthesis of the natural and preternatural worlds, with knowledge a broad continuum ranging from theology to natural philosophy. Argues that the revolutionary attitude that nature was autonomous was not derived from the physical sciences, as most historians believe, but from natural history.

10. Beach, Mark. "Was There a Scientific Lazzaroni?" *Nineteenth-Century American Science: A Reappraisal* (item 32), pp. 115-132.

Denies that the Lazzaroni (nine friends who held positions of power and influence in the American scientific community) acted as a cabal to control American science. Suggests that those out of power created the myth of a conspiratorial group.

11. Beardsley, Edward H. "The American Scientist as Social Activist: Franz Boas, Burt G. Wilder, and the Cause of Racial Justice, 1900-1915." *Isis*, 64 (1973): 50-66.

Reexamines the thesis that the American scientific community was racist during the early twentieth century. Presents two exceptions: an anatomist whose personal experiences led him to reject racism and an anthropologist who viewed generalizations about Black inferiority as examples of the unscientific, unsupported, or amateurish speculations which he wanted exorcised from anthropology.

12. Beaver, Donald de B. "Altruism, Patriotism, and Science: Scientific Journals in the Early Republic." *American Studies*, 12 (1971): 5-19.

Argues that the founders of early American scientific journals acted in the belief that scientific knowledge was basic to national growth. Perceives these men as making self-conscious altruistic and patriotic gestures; editing a scientific journal was thought of as a form of public service.

13. ———. "The American Scientific Community, 1800-1860:
        A Statistical-Historical Study." Ph.D.
        dissertation, Yale University, 1966.

        Focuses on the 138 most productive scientists
as measured by publication in the most representative
scientific journals of the period. Finds that biology
and geology were the most popular fields of research;
specialists contributed the majority of the papers,
although representing a minority of the scientific
community; most of these scientists had some higher
education; the majority were employed either by academic
institutions or government bureaus; most of these highly
productive scientists were linked to each other through
teacher-pupil relationships.

14. ———. "The Smithsonian Origin of the Royal Society
        Catalogue of Scientific Papers." *Science
        Studies*, 2 (1972): 385-393.

        Finds that the index was the result of Joseph
Henry's expansion of the idea for an index of American
scientific papers originated by E.B. Hunt. Observes that
one of the purposes of the index was to demonstrate to
Europeans the extent of American scientific activity.

15. Bell, Whitfield J., Jr. *Early American Science: Needs
        and Opportunities for Study*. Williamsburg:
        Institute of Early American History and Culture,
        1955. ix + 85 pp. Bibliography, Index.

        Surveys the history and historiography of
American science to 1820. Calls for further research in
such areas as the history of individual sciences, science
education, and the diffusion of scientific information.
Includes selective bibliographies for fifty American
scientists. Demonstrates how far the history of American
science has come in the last twenty-five years, but the
absence of responses to some of Bell's suggestions shows
how much work still remains to be done.

16. ———. "The Reverend Mr. Joseph Morgan, an American
        Correspondent of the Royal Society, 1732-1739."
        *Proceedings of the American Philosophical
        Society*, 95 (1951): 254-261.

        Demonstrates the extensive diffusion of high
culture, especially Newtonian science, in the colonies
by examining the knowledge possessed by a country clergy-
man in New Jersey.

17. Bender, Thomas. "Science and the Culture of American
              Communities: The Nineteenth Century." *History
              of Education Quarterly*, 16 (1976): 63-77.

         Asks historians to focus more on science as a
    part of the local culture, especially on the crude em-
    piricism which eventually gave way to the more theoretical,
    nationally or internationally oriented science practiced
    by the professional community. Calls such local community
    science "vernacular."

18. Bland, Larry I. "The Rise of the United States to World
              Scientific Power, 1840-1940." *The History
              Teacher*, 11 (1977): 75-92.

         Attempts to make the rise of American science
    intelligible to general American historians. Identifies
    three major influences on the development of American
    science: professionalization, the restructuring of higher
    education, and the development of mechanisms to allocate
    resources for the support of research. Concludes that
    the current situation did not suddenly materialize as a
    result of World War II, but was the product of over a
    century of evolution.

19. Boller, Paul F., Jr. "The New Science and American
              Thought." *The Gilded Age*. Edited by H. Wayne
              Morgan. Revised edition. Syracuse: Syracuse
              University Press, 1971, pp. 239-257.

         Finds tremendous scientific achievement by
    Americans during the Gilded Age. Argues that the myth of
    the backwardness of American science in basic research
    was the result of the inability of the layman to grasp
    the significance of the work of Gibbs, Rowland, Pickering,
    and many others.

20. Brasch, Frederick E. "The Newtonian Epoch in the American
              Colonies (1680-1783)." *Proceedings of the
              American Antiquarian Society*, n.s., 49 (1939):
              314-332.

         Demonstrates the dominant role of Newtonian
    science in the development of the physical sciences in
    the colonies. Argues that the Newtonian mechanical view
    of nature was an alternative to the Calvinist world view.

21. Bruce, Robert V. "A Statistical Profile of American
                Scientists, 1846-1876." *Nineteenth-Century
                American Science: A Reappraisal* (item 32),
                pp. 63-94.

        Analyzes 477 scientists active during the period
    1846-1876 subsequently included in the *Dictionary of
    American Biography*. Finds that typically they were born
    in New England or the Middle Atlantic States, the sons of
    professionals; college educated; interested in biology,
    chemistry, or the earth sciences; and employed by either
    an educational institution or a government agency.

22. Buck, Peter. *American Science and Modern China, 1876-1936*.
                Cambridge, London, New York: Cambridge University
                Press, 1980. ix + 283 pp. Bibliography, Index.

        Analyzes the introduction of modern science
    into China in an effort to throw light on the changes in
    the organization of American science and its relationship
    to society. Argues that the changing attitudes of Americans
    towards Chinese science reflected American unhappiness
    with the effects of science in their own country. Science
    had not, as had been hoped, reduced social divisions but
    had accentuated the fragmentation along ethnic and class
    lines.

23. Burnham, John C. "Lester Frank Ward as Natural Scientist."
                American Quarterly, 6 (1954): 259-264.

        Concludes that Ward's contributions were not
    significant and his reputation has been inflated.

24. Clark, Harry Hayden. "The Influence of Science on American
                Ideas, from 1775 to 1809." *Transactions of
                the Wisconsin Academy of Sciences, Arts, and
                Letters*, 35 (1943): 305-349.

        Suggests possible influences of scientific
    concepts upon religious, political, humanitarian, and
    educational thought.

25. ———. "The Influence of Science on American Literary
                Criticism, 1860-1910, Including the Vogue of
                Taine." *Transactions of the Wisconsin Academy
                of Sciences, Arts, and Letters*, 44 (1955):
                109-164.

        Argues that Neo-Lamarckian views of the influence
    of the environment and heredity on the individual provided

a framework for literary criticism during the period when
American literature was moving towards realism and
naturalism. Finds that Taine's thesis--a writer was
determined by his race, moment, and milieu and had to be
evaluated within these contexts rather than judged against
some fixed standard--was widely accepted by the end of
the century.

26. Cohen, I. Bernard. *Science and American Society in the
First Century of the Republic*. Columbus: Ohio
State University, 1961. 34 pp. Bibliography.

Presents the nineteenth century as an era of
little scientific progress in the United States, especially
when compared to the great achievements of the colonial
period. The author subsequently revised some of his
conclusions (see item 27).

27. ———. "Science and the Growth of the American Republic."
The Review of Politics, 38 (1976): 359-398.

Concludes that the state of American science
after the Revolutionary War reflected a young, developing
nation. Characterizes the subsequent growth of science
as slow and steady. Revises some of his earlier adverse
conclusions on the state of nineteenth-century American
science (see item 26).

28. Coleman, William. "Science and Symbol in the Turner
Frontier Hypothesis." *American Historical
Review*, 72 (1966): 22-49.

Argues that Turner borrowed concepts from Neo-
Lamarckian evolutionary theory in developing his thesis
of the impact of the environment on the social organism.

29. Cravens, Hamilton. "American Science Comes of Age: An
Institutional Perspective, 1850-1930."
American Studies, 17 (1976): 49-70.

Rejects both the indifference to basic research
and the antebellum institutional development schools.
Argues that internal changes in post-Civil War science--
especially the information explosion--led to a new set of
institutional arrangements, which, in turn, led to the
rapid growth of American science.

30. Daniels, George H. *American Science in the Age of Jackson*. New York and London: Columbia University Press, 1968. ix + 282 pp. Appendices, Index.

   Finds that the leaders of the American scientific community between 1815 and 1845 tended to specialize, had science-related employment, were not preoccupied with the application or practical aspects of science, and were almost as interested in the physical sciences as the biological. Argues that the Baconian philosophy had been embraced by American intellectuals during this period, resulting in a scientific community which believed in the collection of facts, the rejection of theorizing, and the identification of all science with taxonomy. Blames this commitment to empiricism for the lack of American achievement in abstract science. Despite criticism, this book remains highly influential and oft-cited.

31. ———. "Introduction." *Nineteenth-Century American Science: A Reappraisal* (item 32), pp. vii-xv.

   Views this century as a period of flux between colonial dependency and maturity. Warrs historians against the uncritical acceptance of the rhetoric of nineteenth-century scientists as truthful descriptions of the situation.

32. ———, editor. *Nineteenth-Century American Science: A Reappraisal*. Evanston: Northwestern University Press, 1972. xv + 274 pp.

   Provides a reexamination of the historiography, the research role of American scientists, and the relationship between science and technology. Represents the state of the art at the time. Includes items 10, 21, 31, 38, 68, 79, 182, 273, 507, 625, 708, 723.

33. ———. "The Pure-Science Ideal and Democratic Culture." *Science*, 156 (1967): 1699-1705.

   Contends that post-Civil War American scientists switched their justification for research from utility to science for science's sake. Finds that this new justification clashed with the democratic aversion towards public support of an elite for purposes other than that of the general good.

34. ————. *Science in American Society: A Social History*.
         New York: Alfred A. Knopf, 1971. xii + 390 pp.
         Bibliography, Index.

        Attempts to provide an overview of the develop-
    ment of science within the American cultural context. The
    author has publicly admitted that portions of this book
    represent verbatim use of other scholars' work without
    proper acknowledgment.

35. Davis, Richard Beale. "Science and Technology, Including
         Agriculture." *Intellectual Life in the Colonial
         South, 1585-1763*. Knoxville: The University of
         Tennessee Press, 1978, 2:805-987, 1073-1112.

        Concludes that Southern scientific activity was
    heavily dependent upon European inspiration. Finds less
    interest in astronomy than among New Englanders but wide-
    spread concern with natural history. Sees slavery as
    having little impact upon Southern scientific activity.

36. Dolby, R.G.A. "The Transmission of Two New Scientific
         Disciplines from Europe to North America in the
         Late Nineteenth Century." *Annals of Science*,
         34 (1977): 287-310.

        Finds that both experimental psychology and
    physical chemistry were transmitted by Americans trained
    in German teaching laboratories. Shows that physical
    chemistry was transmitted whole because there was no
    prior research tradition in America for this subject and,
    hence, no conflict. Observes that in contrast, experimental
    psychology had to contend with an established tradition and
    could only be transferred selectively.

37. Dupree, A. Hunter. "The History of American Science--A
         Field Finds Itself." *American Historical
         Review*, 71 (1966): 863-874.

        Defends the study of the history of American
    science as part of interdisciplinary studies of American
    civilization. Examines the impact of Arthur M. Schlesinger,
    Sr.'s vision of science and technology as central themes
    in American history. Contrasts Schlesinger's attitude
    that these subjects were the purview of American historians
    with George Sarton's insistence that the historians of
    science have expertise in science.

38. ————. "The Measuring Behavior of Americans." *Nineteenth-Century American Science: A Reappraisal* (item 32), pp. 22-37.

    Defines science as a measurement system and measurement as a method of making information about the environment transmittable and comparable. Identifies four measurement systems in nineteenth-century America: rectangular grids, spherical coordinates, biological classifications, and geological dating.

39. Elliott, Clark A. "The American Scientist in Antebellum Society: A Quantitative View." *Social Studies of Science*, 5 (1975): 93-108.

    Examines 538 individuals engaged in scientific activity between 1800 and 1863. Finds that they generally came from urban, upper middle class or upper class, commercial or professional families, were college graduates, and had science-related jobs. Compares their profile to other occupations and concludes that the overall pattern is much closer to literary men than inventors.

40. ————. *Biographical Dictionary of American Science: The Seventeenth Through the Nineteenth Centuries.* Westport, Conn. and London: Greenwood Press, 1979. xvii + 360 pp. Appendices, Index.

    Provides six hundred biographical sketches and cross-references to another three hundred scientists. Includes references to manuscript holdings.

41. Fermi, Laura. *Illustrious Immigrants: The Intellectual Migration from Europe, 1930-41.* 2d edition. Chicago and London: The University of Chicago Press, 1971. xii + 431 pp. Index.

    Summarizes the experiences of nineteen hundred immigrants. Finds that most were Germans or Austrians arriving after 1937. Describes the contributions and achievements of the immigrants in a number of specific fields, including psychoanalysis, atomic physics, mathematics, astronomy, molecular biology, and the social sciences. Contends that most made the adjustments necessary for fruitful careers in their new homelands.

42. Fleming, Donald. "Science in Australia, Canada, and the United States: Some Comparative Remarks." *Proceedings of the Tenth International Congress of the History of Science.* Paris: Hermann, 1964, 1:179-196.

Compares and contrasts the developments in
three nations which shared a scientific heritage and
faced common tasks. Finds that the American Revolution
forced Americans to restructure the institutional life
of their science and construct substitutes for English
institutions.

43. Gerbi, Antonello. *The Dispute of the New World: The
History of a Polemic, 1750-1900*. Translated
by Jeremy Moyle. Revised edition. Pittsburgh:
University of Pittsburgh Press, 1973. xviii +
700 pp. Bibliography, Index.

Traces the rise and decline of the theory of
the inferiority of the men, fauna, and climate of the New
World to that of the Old World. Contends that Buffon's
exposition was simply a coherent presentation and synthesis
of ideas unsystematically but widely present in the accounts
of travellers, missionaries, and naturalists. Finds that
the polemic peaked with the attacks by Hegel. Argues
that as scientific knowledge of the New World increased
during the nineteenth century, the focus of the accusa-
tions shifted from the alleged inferiority of the natural
world to the immaturity of the society and culture present
in the nations of the Americas. Sees Mrs. Trollope's
attacks as an example of this new version of the theory.
Contains a very extensive bibliography.

44. Gillispie, Charles Coulston, editor. *The Dictionary of
Scientific Biography*. New York: Charles
Scribner's Sons, 1970-1980.

Offers entries of very uneven quality, ranging
from trivial recitations of facts to original analyses of
the lives and contributions of the subjects. Despite its
weaknesses, this should be the first reference consulted
for biographical information.

45. Goetzmann, William H. *Exploration & Empire: The Explorer
and the Scientist in the Winning of the American
West*. New York: Alfred A. Knopf, 1966. xxvi
+ 656 + xviii pp. Bibliography, Index.

Examines the role of Western exploration in
the development of the nation, specifically the impact of
this exploration upon science, scientific institutions,
and public policy towards the West. Divides the exploration into three phases: 1805-1845, 1845-1860, and 1860
through the end of the century. Each phase was marked by

progressively less emphasis upon geographical discovery
and increasing concern with the systematic cataloging of
natural resources and scientific information.

46. ————. "Paradigm Lost." *The Sciences in the American
          Context: New Perspectives* (item 84), pp. 21-34.

        Asserts that a paradigm shift--the Second Great
Age of Discovery, stretching from 1600 to 1900--has been
ignored by historians of science. Argues that the
characteristic system building of this period was due to
the tremendous information explosion resulting from the
explorations and surveys conducted during these years.

47. Greene, John C. "American Science Comes of Age, 1780-
          1820." *The Journal of American History*, 55
          (1968): 22-41.

        Finds that American science remained a branch
of British science long past the declaration of political
independence. Concludes that it was not until after the
War of 1812 that rising nationalism led to the establishment of the institutions which were the prerequisites for
an independent American scientific community.

48. Hall, Courtney Robert. *A Scientist in the Early Republic:
          Samuel Lantham Mitchill, 1764-1831.* New York:
          Columbia University Press, 1934. Reprint.
          New York: Russell and Russell, 1967. vii +
          162 pp. Appendices, Bibliography, Index.

        Provides an overview of one of the pioneers in
the natural sciences during the Early National Period.
Discusses his contributions to geology, zoology (especially
ichthyology), and botany. Argues that the driving force
in his life was his nationalism, and regards his scientific
activities and his political career as two sides of this
nationalistic commitment.

49. Hall, Michael G. "Renaissance Science in Puritan New
          England." *Aspects of the Renaissance: A
          Symposium.* Edited by Archibald R. Lewis.
          Austin and London: University of Texas Press,
          1967, pp. 123-136.

        Argues that the Renaissance attributes of
Puritan science were not fully displaced in New England
until about 1700. Offers the evolution of astronomical
theory from the geocentric view to the Keplarian theory,
traced through almanacs, as a case study.

50. Hindle, Brooke, editor. *Early American Science*. New
York: Science History Publications, 1976.
xiv + 213 pp.

Reprints 22 *Isis* articles dealing with American
science prior to 1820. Includes items 53, 88, 90, 97,
98, 231, 386, 387, 434, 501.

51. ————. "The Historiography of Science in the Middle
Colonies and the Early Middle States."
*Proceedings of the American Philosophical
Society*, 108 (1964): 158-161.

Maintains that this geographical unit is
irrelevant for the purpose of analyzing the history of
science during this period. New Jersey served as an
insulator between New York City, which was linked to New
England, and Philadelphia, which looked southward, rather
than as a link between the two cities.

52. ————. *The Pursuit of Science in Revolutionary America,
1735-1789*. Chapel Hill: The University of
North Carolina Press, 1956. Reprint. New
York: W.W. Norton & Company, Inc., 1974.
xi + 410 pp. Bibliography, Index.

Describes the naturalists, physicians, natural
philosophers, clergymen, and teachers who made up the
colonial scientific community and delineates their
connections with their European colleagues. Discusses
the establishment of the American Philosophical Society.
Surveys colonial accomplishments in natural history, the
physical sciences, and technology, with particular emphasis
on the observations of the Transit of Venus. Argues that
the Revolutionary generation was confident that science
would flourish in the Republic; in turn, science would
prove fundamental for the progress of the nation.

53. ————. "The Quaker Background and Science in Colonial
Philadelphia." *Isis*, 46 (1955): 243-250.

Argues that the empirical, rational, worldly
approach to knowledge produced a favorable atmosphere for
science, although the peripheral or former Quakers, and
non-Quakers living in the Quaker environment, were the
most supportive.

54. Hollinger, David A.   "Science and Anarchy: Walter Lippmann's
        Drift and Mastery." *American Quarterly*, 29
        (1977): 463-475.

        Argues that Lippmann saw science as a method
    and a spirit which could be applied by society to establish
    a new stability during an age of transition. Among the
    scientist's attributes were a sense of intersubjectivity
    and personal commitment.

55. Horowitz, Helen L.   "Animal and Man in the New York
        Zoological Park." *New York History*, 56
        (1975): 426-455.

        Evaluates the development of the New York Zoo
    as a reflection of the founders' attitudes towards wild
    animals and inferior types of men. Contends that the zoo
    was an example of the environments established by America's
    elite during the great social and environmental changes
    of the early twentieth century to control animals and
    humans.

56. Jaffe, Bernard. *Men of Science in America: The Story of
        American Science Told Through the Lives and
        Achievements of Twenty Outstanding Men from
        Earliest Colonial Times to the Present Day.*
        Revised edition. New York: Simon and Schuster,
        1958. Reprint. New York: Arno Press, 1980.
        xli + 715 pp. Bibliography, Index.

        Utilizes individual biographies as starting
    points for the analysis of broader themes. Represents
    one of only a handful of attempts to present the history
    of American science from the colonial era to the present
    in one volume. This was the standard text in the 1940s
    and 1950s (in its first edition), but it is now judged
    to be dated and unreliable.

57. Johnson, Thomas Cary, Jr. *Scientific Interests in the
        Old South.* New York: D. Appleton-Century
        Company, 1936. Reprint. Wilmington and
        London: Scholarly Resources, Inc., 1973.
        vi + 214 pp. Index.

        Defends the South against charges of indifference
    to science during the antebellum era. Catalogs the forms
    of scientific activity and evidence of widespread interest
    in science on a popular level. Discusses the presence of
    science in female seminaries and in magazines oriented
    towards women. Concludes that the inability to form

cooperative ventures and the disdain of specialization
were the major factors in restricting the contributions
of the South to the progress of science.

58. Kargon, Robert H. "Temple to Science: Cooperative Research
and the Birth of the California Institute of
Technology." *Historical Studies in the Physical
Sciences*, 8 (1977): 3-31.

Treats Caltech as the end product of George E.
Hale's aspirations to establish an institutional setting
for cooperative interdisciplinary scientific research.
Argues that Hale drew his inspiration from his experiences
in astrophysics.

59. Kevles, Daniel J. "'Into Hostile Political Camps': The
Reorganization of International Science in
World War I." *Isis*, 62 (1971): 47-60.

Traces Hale's efforts to reorganize international
science during World War I, which culminated in the establishment of the International Research Council. Identifies
Hale's objectives as the exclusion of the Central Powers
from international science and stronger international
cooperation among the Allies and neutrals.

60. ————, Jeffrey L. Sturchio, and P. Thomas Carroll.
"The Sciences in America, Circa 1880."
Science, 209 (1980): 27-32.

Finds that localism still characterized most
American scientific activity although the transition to
a national community had begun. Argues that the strength
of American research lay in observation, experimentation,
measurement, and the development and/or exploitation of
new instrumentation.

61. Klein, Aaron E. *The Hidden Contributors: Black Scientists
and Inventors in America*. Garden City, N.Y.:
Doubleday & Company, Inc., 1971. xiii + 203
pp. Index.

Presents twelve examples. Argues that past
segregation policies and racist attitudes have resulted
in a waste of human resources.

62. Kohlstedt, Sally Gregory. *The Formation of the American
            Scientific Community: The American Association
            for the Advancement of Science, 1848-1860.*
            Urbana, Chicago, London: University of Illinois
            Press, 1972. xiii + 264 pp. Bibliography,
            Index, Appendix.

Examines the early history of the AAAS in light
of the conflicting objectives of the popularization of
science and open access versus the advancement of scientific
research and professionalization. Finds that the professional outlook increasingly dominated the AAAS during
the 1850s. Includes a profile of the membership, a list
of members, and biographical sketches of the leadership.

63. ————. *"Science*: The Struggle for Survival, 1880-1894."
            Science, 209 (1980): 33-42.

Argues that *Science* survived its infancy because
it deferred to the interests and priorities of the leaders
of the scientific community, quickly discovered a format
and content which was attractive to the scientific
community, and was able to obtain at least minimal
funding. Identifies the lack of professional science
writers as one of the journal's major problems during
this period.

64. Kopp, Carolyn. "The Origins of the American Scientific
            Debate over Fallout Hazards." *Social Studies
            of Science*, 9 (1979): 403-422.

Illustrates the thesis that the social interests
which often supply the context of scientific disputes are
multidimensional and interactive. Argues that the social
roots of this debate were the methodological and conceptual
differences between geneticists and other scientists, the
institutional differences between academics and the Atomic
Energy Commission, personal differences between scientists
and the chairman of the AEC, and political differences
between liberal and conservative scientists.

65. Levenstein, Harvey. "The New England Kitchen and the
            Origins of Modern American Eating Habits."
            American Quarterly, 32 (1980): 369-386.

Attributes the failure of the late nineteenth-
century attempt to reform working-class eating habits to
a lack of understanding of the psychological role of
certain foods in this class's lifestyle. Finds that
this reform movement was based on little scientific
research and would have resulted in nutritional disaster.

66. Leverette, William E., Jr. "E.L. Youman's Crusade for
             Scientific Autonomy and Respectability."
             American Quarterly, 17 (1965): 12-32.

        Describes the campaign carried out in the pages
    of the *Popular Science Monthly* from 1872-1887 to persuade
    America that scientists' search for truth deserved a
    respect and social status which to date had not been
    granted.

67. Lindsay, G. Carroll. "George Brown Goode." *Keepers of
             the Past*. Edited by Clifford L. Lord. Chapel
             Hill: The University of North Carolina Press,
             1965, pp. 127-140.

        Discusses Goode's contributions to the evolution
    of the museum function and the increasing emphasis upon
    the arts, industries, and history in the Smithsonian
    Institution.

68. Lurie, Edward. "The History of Science in America:
             Development and New Directions." *Nineteenth-
             Century American Science: A Reappraisal* (item
             32), pp. 3-21.

        Records the development of the history of
    American science over the previous twenty-five years.
    Contends that this specialty has achieved comparable
    status with general American history and the history
    of science.

69. ———. "An Interpretation of Science in the Nineteenth
             Century: A Study in History and Historiography."
             Journal of World History, 8 (1965): 681-706.

        Attacks the then-prevalent position that American
    scientific accomplishments during the nineteenth century
    were insignificant. Argues that the nineteenth century
    was the essential period of institution building and
    organization which must precede research. Surveys the
    need for historical research.

70. ———. "Science in American Thought." *Journal of
             World History*, 8 (1965): 638-665.

        Argues that American culture was transformed
    from a "popularist, expansive, social democratic" form
    to "an ordered, professionalized, institutional and
    republican" form during the years 1840-1880. Sees
    science as an intimate part of that culture, valued for
    its ability to morally uplift the citizen and transform
    the environment.

71. McAllister, Ethel M. *Amos Eaton: Scientist and Educator*.
        Philadelphia: University of Pennsylvania Press,
        1941. xiv + 587 pp. Bibliography, Index.

        Discusses Eaton's activities in geology,
    mineralogy, zoology, chemistry, and botany. Argues that
    Eaton was a strong advocate of artificial classification
    schemes in botany because they were easily learned by the
    beginner. Examines Eaton's pedagogical theory which
    emphasized learning by doing and the importance of
    scientific knowledge and discusses the institutional
    expression of that theory, Rennselaer Polytechnic Institute.

72. Martin, Jean-Pierre. "Edwards' Epistemology and the New
        Science." *Early American Literature*, 7 (1973):
        247-255.

        Argues that Jonathan Edwards's epistemology did
    not incorporate the developments in eighteenth-century
    science.

73. Meltsner, Arnold J. "The Communication of Scientific
        Information to the Wider Public: The Case of
        Seismology in California." *Minerva*, 17 (1979):
        331-354.

        Identifies two forms of the suppression of
    scientific knowledge in popular media: external suppression
    by individuals wanting to play down potential dangers,
    and self-suppression by the scientific community which
    feared creating panic. Finds that since scientists have
    little training in the dissemination of information to
    the public, they frequently do it poorly.

74. Mendelsohn, Everett. "Science in America: The Twentieth
        Century." *Paths of American Thought*. Edited
        by Arthur M. Schlesinger, Jr. and Morton
        White. Boston: Houghton Mifflin Company,
        1963, pp. 432-445.

        Outlines the quantitative and qualitative
    growth of American science during this century.

75. Morantz, Regina Markell. "The Scientist as Sex Crusader:
        Alfred C. Kinsey and American Culture."
        American Quarterly, 29 (1977): 563-589.

        Contends that Kinsey was not a social revolu-
    tionary despite his questioning of society's sexual mores.
    His revolt against American values was very narrow, limited,
    and based on his observations of biological activity.

76. Nichols, Roger L. and Patrick L. Halley. *Stephen Long and American Frontier Exploration*. Newark: University of Delaware Press, 1980. 276 pp. Appendix, Bibliography, Index.

    Concentrates on Long's activities as a promoter and leader of frontier exploration because there is insufficient material for a complete biography. Credits Long with establishing the precedent of using civilian scientists on military expeditions. Finds that Long measured success in terms of miles travelled, resulting in unrealistic goals and limited opportunities for scientific observations. Questions the competence of his leadership.

77. Post, Robert. "Science, Public Policy, and Popular Precepts: Alexander Dallas Bache and Alfred Beach as Symbolic Adversaries." *The Sciences in the American Context: New Perspectives* (item 84), pp. 77-98.

    Compares and contrasts the attitudes of Bache, representative of the professional scientific community, and Beach, editor of the *Scientific American*, who was the spokesman for the more democratic approach to science. Praises Bache for his skill at courting popular support through public relations and exploiting the political system for the sake of scientific research.

78. Randel, William Peirce. "Huxley in America." *Proceedings of the American Philosophical Society*, 114 (1970): 73-99.

    Describes the 1876 lecture tour of T.H. Huxley, British naturalist and propagandist for Darwinian evolution, which was highlighted by an address at the inaugural ceremonies at Johns Hopkins University.

79. Reingold, Nathan. "American Indifference to Basic Research: A Reappraisal." *Nineteenth-Century American Science* (item 32), pp. 38-62.

    Questions one of the basic tenets of the historiography of American science (see item 89). Argues that the indifference school has substituted criticism of the American scientific community for failing to meet ideal goals established by the historian for the study of the actual accomplishments of the community within their historical context. Calls for more quantitative, comparative data to delineate the real differences and similarities

in the conduct and support of science in the United
States and Europe.

80. ———. "National Aspirations and Local Purposes."
        Transactions of the Kansas Academy of Sciences,
        71 (1968): 235-246.

        Warns of the dangers of selective citation and
simplistic formulations in understanding historical situations. Analyzes the activities of James McKeen Cattell in
the American Association for the Advancement of Science
and George Ellery Hale in the National Academy of Sciences
to demonstrate the limitations of simple, black and white
interpretations.

81. ———. "Reflections on Two Hundred Years of Science in
        the United States." *Nature*, 262 (1976): 9-13.

        Provides a brief survey of the historiographic
problems and possible future themes in the history of
American science.

82. ———, editor. *Science in America Since 1820.* New York:
        Science History Publications, 1976. iii +
        334 pp.

        Reprints twenty-one articles from *Isis.* Includes
items 11, 59, 167, 202, 207, 238, 239, 256, 270, 308, 320,
332, 341, 350, 388, 446, 512, 513, 643, 692.

83. ———, editor. *Science in Nineteenth Century America:
        A Documentary History.* New York: Hill and
        Wang, 1964. Reprint. New York: Octagon Books,
        1979. xii + 339 pp. Index.

        Provides an overview of the major institutional
and intellectual developments. Analyzes the strengths
and weaknesses of American science on its own terms.
Introduces the concept of the science of geography--
subdivided into natural history and geophysics--as an
organizing principle in understanding American science.

84. ———, editor. *The Sciences in the American Context:
        New Perspectives.* Washington, D.C.: Smithsonian
        Institution Press, 1979. 399 pp.

        Contains fifteen papers prepared in conjunction
with the Bicentennial. Includes item 46, 77, 81, 91, 176,
180, 268, 344, 352, 413, 456, 523, 713, 733.

85. ———. "The Scientist as Troubled American." *Journal
        of the American Industrial Hygiene Association*,
        40 (1979): 1107-1113.

    Links the complaints of American scientists
regarding society's lack of support for basic research to
their perception that Americans do not grant scientists
the deference the scientific community believes they
deserve.

86. Richmond, Phyllis Allen. "The Nineteenth Century American
        Physician as a Research Scientist." *Inter-
        national Record of Medicine*, 171 (1958):
        492-506.

    Claims that there was much more basic research
in the United States than is usually assumed and offers
examples. Finds that almost half of the American
scientists who made significant contributions in basic
research were trained as physicians.

87. Rosenberg, Charles E. *No Other Gods: On Science and
        American Social Thought*. Baltimore and London:
        The Johns Hopkins University Press, 1976.
        xiii + 273 pp. Bibliography, Index.

    Combines twelve previously published essays
with one new piece offering case studies illuminating the
relationship between science and society. Delineates
four aspects of that relationship: science as a motivator
of behavior; science as a supplier of social images; the
role of the professional scientist in society; and the
impact of society upon the scientist. Argues that the
successful scientist had to learn how to utilize a
variety of constituencies, audiences, and benefactors.
Recognizes that society borrowed the vocabulary of
science to explain social change and shape social values.

88. Schlesinger, Arthur M., Sr. "An American Historian Looks
        at Science and Technology." *Isis*, 36 (1946):
        162-166.

    Pleads for a history of science which would
trace the connections between science and society, as
well as identify the characteristics which distinguish
American from world science. Advocates the theory that
social need leads to technological innovation.

89. Shryock, Richard Harrison. "American Indifference to
             Basic Science During the Nineteenth Century."
             *Archives Internationales d'Histoire des
             Sciences*, 28 (1948): 50-65.

       Blames America's lag in basic research on the
    utilitarian prejudices of American society. Represents a
    major interpretation of the history of science in America.
    The weaknesses of the argument are exposed in item 79.

90. ————. "The Need for Studies in the History of American
             Science." *Isis*, 35 (1944): 10-13.

       Calls for more history of science in general
    American history courses. Complains that American
    historians are ignorant of the history of science despite
    science's role in modern society and culture.

91. Sinclair, Bruce. "Americans Abroad: Science and Cultural
             Nationalism in the Early Nineteenth Century."
             *The Sciences in the American Context: New
             Perspectives* (item 84), pp. 35-53.

       Focuses on the American scientists who travelled
    to Europe in the 1830s. Finds that this experience en-
    hanced their reputations, developed their international
    connections, made them more aware of the limitations of
    American science, and strengthened their desire to improve
    the state of that science.

92. Sloan, Douglas. "Science in New York City, 1867-1907."
             Isis, 71 (1980): 35-76.

       Analyzes the changing relationships among
    institutions concerned with natural history as New York
    (and American) science was transformed from community
    oriented to nationally oriented. Utilizes the career of
    John S. Newberry to demonstrate these changes.

93. Sokal, Michael M. "*Science* and James McKeen Cattell,
             1894-1945." *Science*, 209 (1980): 43-52.

       Finds that despite *Science*'s status as the
    official journal of the American Association for the
    Advancement of Science, its pages during these years
    reflected the attitude of its editor, J.M. Cattell, not
    the membership. At his death the AAAS had difficulty
    defining its relationship to its own journal.

94. Stearns, Raymond Phineas. *Science in the British Colonies of America*. Urbana: University of Illinois Press, 1970. xx + 760 pp. Appendices, Bibliography, Index.

Emphasizes the role of the Royal Society of London in the promotion of colonial scientific activity during the seventeenth century. Finds that individual English scientists filled the same role during the eighteenth century as the Royal Society went into decline. Uncovers a considerable amount of new information. The bibliography and footnotes serve as a guide to the literature and manuscript collections in the field.

95. Struik, Dirk J. *Yankee Science in the Making*. Revised edition. New York: Collier Books, 1962. 544 pp. Bibliography, Index.

Represents an early attempt at developing a social history of science and technology. Limits itself to New England prior to the Civil War because of the presence of a relatively homogeneous and stable population. Divides the chronology into Federalist and Jacksonian eras. During the former period, science developed in the coastal towns and reflected the interests of the mercantile class, while technology developed completely independently among the interior farmers. The latter period was marked by mass interest in science, evidenced by the lyceum movement, as well as closer ties between science and technology as the mercantile interests gave way to the industrialists.

96. Thackray, Arnold. "Reflections on the Decline of Science in America and on Some of Its Causes." *Science*, 173 (1971): 27-31.

Views the relationship between science and society as precarious and easily disfunctioned. Surveys this relationship and draws parallels between the problems that face the contemporary American scientific community and those of scientific communities of earlier eras and different nations.

97. Tolles, Frederick B. "Philadelphia's First Scientist: James Logan." *Isis*, 47 (1956): 20-30.

Sketches the life and intellectual contributions of the leading intellectual and political figure in Pennsylvania during the first half of the eighteenth century.

98. Tucker, Leonard. "President Thomas Clap of Yale College:
    Another 'Founding Father' of American Science."
    Isis, 52 (1961): 55-77.

    Illustrates the integral role science played
in the lives of colonial intellectuals, even those whose
theological orientation was extremely conservative.

99. Van Tassel, David D., and Michael G. Hall, editors.
    Science and Society in the United States.
    Homewood, Illinois: The Dorsey Press, 1966.
    vi + 360 pp. Bibliography, Chronology,
    Glossary, and Index.

    Surveys the relationship between scientists,
engineers, their organizations, and the larger society.
Includes items 183, 276, 324, 345, 715, 746.

100. Walsh, John. "*Science* in transition, 1946 to 1962."
    Science, 209 (1980): 52-57.

    Contends that the journal floundered until the
mid-50s, when it hired a professional staff and determined
that its primary appeal was to academic scientists.

101. Washburn, Wilcomb E. "The Influence of the Smithsonian
    Institution on Intellectual Life in Mid-Nine-
    teenth-Century Washington." *Records of the
    Columbia Historical Society*, 63-65 (1966):
    96-121.

    Claims that the Smithsonian made Washington a
cultural center by sponsoring lectures and attracting
out-of-town scholars and scientists.

102. ————. "Joseph Henry's Conception of the Purpose of
    the Smithsonian Institution." *A Cabinet of
    Curiosities: Five Episodes in the Evolution
    of American Museums* (item 103), pp. 106-166.

    Sees Henry's commitment to the increase of
knowledge resulting in the subordination of the museum
function to the support of research and publication.

103. Whitehill, Walter Muir, editor. *A Cabinet of Curiosities:
    Five Episodes in the Evolution of American
    Museums.* Charlottesville: The University Press
    of Virginia, 1967. xii + 166 pp.

    Contains items 102, 191, 212, 235, 635.

104. Whitford, Kathryn, and Philip Whitford. "Timothy Dwight's Place in Eighteenth-Century American Science." *Proceedings of the American Philosophical Society*, 114 (1970): 60-71.

Argues that this president of Yale was a gentleman-scientist in the same mold as Thomas Jefferson.

105. Wilson, Mitchell. *American Science and Invention: A Pictorial History*. New York: Bonanza Books, 1960. x + 437 pp. Bibliography, Index.

Surveys the interplay among American science, technology, and society, while showing considerable interest in the personalities of the scientists and inventors. Contains some magnificent illustrations. This is a coffee-table book for the general reader.

GENERAL TECHNOLOGY

106. Akin, William E. *Technocracy and the American Dream: The Technocratic Movement, 1900-1941*. Berkeley: University of California Press, 1977. xv + 227 pp. Bibliography, Index.

Finds the roots of the Technocrat movement in the Progressive drive for efficiency, expertise, and planning; the engineer's search for a professional identity; and the theories of Thorstein Veblen. Discusses the movement's critique of capitalism and the role of technology in modern society. Concentrates on the movement during the 1930s. Argues that it failed because it was unable to develop a political philosophy acceptable to the American public.

107. Borut, Michael. "The *Scientific American* in Nineteenth Century America." Ph.D. dissertation, New York University, 1977.

Emphasizes the antebellum period, focusing on the role of editors and publishers in establishing the policy of the journal. Argues that the journal and its patent agency were important auxiliaries in the progress of the Industrial Revolution in the United States. Links the self-help policy for mechanics advocated in the journal to the philosophy of the Mechanics Mutual Protection, a conservative labor organization which advocated self-help through learning the basic scientific principles which could be applied on the job.

108. Burlingame, Roger. *March of the Iron Men: A Social History
              of Union Through Invention*. New York: Charles
              Scribner's Sons, 1938. Reprint. New York:
              Arno Press, 1976. xvi + 500 pp. Bibliography,
              Appendix, Index.

         Claims that Americans responded to their hostile
    environment by turning to technology for assistance.
    Organizes material according to a generally chronological
    pattern. This book and its companion volume (item 109)
    are not reliable guides to social history.

109. ————. *Engines of Democracy: Invention and Society in
              Mature America*. New York: Charles Scribner's
              Sons, 1940. Reprint. New York: Arno Press,
              1976. xviii + 606 pp. Bibliography, Index.

         This is a continuation of item 108. Utilizes
    a thematic approach. Appears generally optimistic about
    the impact of technology on the human race.

110. Chapin, Seymour L. "Patent Interferences and the History
              of Technology: A High-Flying Example." *Tech-
              nology and Culture*, 12 (1971): 414-446.

         Utilizes the priority dispute over the invention
    of the automatic pressurizing system for aircraft to
    illustrate the value of patent interference records for
    the history of technology. Claims that these records
    provide an index to the state of the art, glimpses into
    inventors' motivations, and dates of the conceptualization
    and reduction to actual practice of an innovation.

111. Daniels, George H. "The Big Questions in the History of
              American Technology." *Technology and Culture*,
              11 (1970): 1-21.

         Identifies the big questions as those dealing
    with technology as a social phenomenon. Offers as
    examples the relationship between technological change
    and social change, the origins of the American system of
    manufacturing, and the source of the widespread diffusion
    of technical skills among the American population.

*   Davis, Richard Beale. "Science and Technology, Including
              Agriculture." *Intellectual Life in the Colonial
              South, 1585-1763*. Cited above as item 35.

         Argues that technology was less dependent upon
    European inspiration than science because of the radical

differences between European and Southern colonial environments, labor supplies, and economic regulations.

112. Ferguson, Eugene S. "The American-ness of American Technology." *Technology and Culture*, 20 (1979): 3-24.

    Contends that America's technology is viewed by most of the world as its most visible and compelling feature. Argues that there are connections between the distinctive qualities of American technology and the central dream of democracy of sharing the good life, the missionary zeal of Americans to proclaim the benefits of their system, and the obvious interest of Americans in machines. Rejects the thesis that innovation can be explained in exclusively economic terms.

113. ————. "On the Origin and Development of American Mechanical 'Know-How.'" *Midcontinent American Studies Journal*, 3 (1962): 3-15.

    Contends that the most important factor in the development of American technology was the accumulation and dissemination of technical knowledge, not any single inventor or particular innovation. Includes among the sources of technical knowledge technical literature, specimens of English machinery, immigrant skilled craftsmen, and reports by American travellers to Europe.

114. Fisher, Marvin. *Workshops in the Wilderness: The European Responses to American Industrialization, 1830-1860*. New York: Oxford University Press, 1967. ix + 238 pp. Bibliography, Index.

    Discusses the antebellum industrialization of the United States as seen through the eyes of European visitors. Focuses on the contradictory visions of the United States as garden and workshop. Argues that Europeans detected unique characteristics to America's industrialization, reflecting the wealth of natural resources, character of the people, and freedom from class and craft restrictions.

115. Fox, Frank W. "The Genesis of American Technology, 1790-1860: An Essay in Long-range Perspectives." *American Studies*, 17 (1976): 29-48.

    Demonstrates the limitations in the previous historiography of American technology. Argues that it was the structure of American society which led to the

distinctiveness of American technology. Characterizes
American society as individualistic, egalitarian, am-
bitious, and open-ended; these are qualities which made
Americans very supportive of innovation.

116. Hindle, Brooke. *Technology in Early America: Needs and
              Opportunities for Study*. Chapel Hill: The
              University of North Carolina Press, 1966.
              xix + 145 pp. Bibliography, Index.

        Contends that the role of science in the devel-
opment of technology during this period was very limited.
Discusses the use of artifacts in pursuing the history
of technology. The bibliography is dated but still
useful, especially in its consideration of contemporary
surveys, histories, textbooks, and manuals.

117. Hughes, Thomas Parke. *Elmer Sperry: Inventor and Engineer*.
              Baltimore and London: The Johns Hopkins
              University Press, 1971. xvii + 348 pp.
              Appendices, Index.

        Views the career of Sperry as a microcosm of
the history of technology in the United States as technology evolved from heroic independent invention to
organized research and development. Presents Sperry as
an important transitional figure in the transformation of
the United States into an industrial nation whose greatest
resource was its high technology. Emphasizes the increasing dependence of technologists upon the military
during and after World War I, and the resulting birth of
the military-industrial complex.

118. Hunter, Louis C. *A History of Industrial Power in the
              United States, 1780-1930*. Volume I: *Waterpower
              in the Century of the Steam Engine*. Charlottes-
              ville: University Press of Virginia, 1979.
              xxv + 606 pp. Appendices, Index.

        Studies the chief generator of power in ante-
bellum America within its social and physical context.
Analyzes both the technology of water wheels and turbines
and the management of water resources. Discusses the
limitations imposed upon the use of waterpower by law
(water rights) and geographical determinants. Traces
the role of waterpower in the industrialization of the
United States, examining the impact of rural mills and
factory villages. Blames the declining use of waterpower
on its unreliability and inflexibility; as the economy
became more nationalized and integrated, industry sought
alternative energy forms.

119. ————. "Waterpower in the Century of the Steam Engine."
*America's Wooden Age: Aspects of its Early
Technology*. Edited by Brooke Hindle. Tarrytown: Sleepy Hollow Restorations, 1975, pp. 160-
192.

Argues that the role of the steam engine in
early industrialization has been exaggerated. Claims
that Americans used steam power only as a last resort
because waterpower was so abundant, inexpensive, and fit
the needs of the small, local industries so characteristic
of the nation, especially near the frontier.

120. Inouye, Arlene, and Charles Süsskind. "'Technological
Trends and National Policy,' 1937: The First
Modern Technology Assessment." *Technology and
Culture*, 18 (1977): 593-621.

Evaluates an attempt by technical experts to
predict possible applications of technology to meet the
problem of unemployment. Fails to fully illuminate the
New Deal context of the assessment.

121. Kasson, John F. *Civilizing the Machine: Technology and
Republican Values in America, 1776-1900*. New
York: Grossman Publishers, 1976. xiv + 274 pp.
Index.

Studies the compatibility and interaction of
republican ideals with the process of industrialization.
Traces the relationship from the initial fears that
technology would produce undesirable social consequences,
through the mid-nineteenth-century position that technology
promised economic independence, social cohesion, and
public virtue, to the return of unease at the end of the
century. Views the utopian novels of the late nineteenth
century as attempts to synthesize technology and re-
publicanism during an era of social unrest. Attacks
Kouwenhoven's thesis regarding the aesthetics of
machinery (see item 123).

122. Kilgour, Frederick G. "Technological Innovation in the
United States." *Journal of World History*, 8
(1965): 742-767.

Claims that America's greatest contributions
have been in the field of production technology, including
mass production, interchangeable parts, the assembly
line, and scientific management.

123. Kouwenhoven, John A. *Made in America: The Arts in Modern Civilization*. New York: Doubleday and Company, 1948. Reprint. New York: Octagon Books, 1975. xv + 303 pp. Bibliography, Index.

Identifies two forms of art in America: a diluted form of European art, divorced from American civilization, functioning as high culture, and a vernacular art, rooted in the civilization, and drawing upon technology for inspiration. Characterizes American machines and hence the aesthetic values of the vernacular art as light, simple, rough, and devoid of ornate decoration. Contends that where the propaganda of high culture was excluded, the vernacular aesthetic values have flourished in a variety of forms, such as furniture, the performing arts, literature, and painting.

124. Layton, Edwin T., Jr., editor. *Technology and Social Change in America*. New York: Harper and Row, 1973. vii + 181 pp. Bibliography.

Reprints ten articles dealing with either the origins and nature of technological change, or the impact of technology upon society. Includes items 113, 129, 288, 755, 774, 778, 801.

125. McCullough, David. *The Great Bridge*. New York: Simon and Schuster, 1972. 636 pp. Appendix, Bibliography, Index.

Examines the history of the construction of the Brooklyn Bridge (1867-1883), the most impressive American engineering feat of its time. Focuses on the efforts of John A. Roebling, who designed the bridge, and his son Washington, who supervised construction after his father's death in 1869. Examines both the technical problems in the construction and the social and political contexts within which the Roeblings had to operate. Studies the history of the medical knowledge of the "bends," which represented a major problem for this project during the digging of the underwater foundations.

126. Marx, Leo. *The Machine in the Garden: Technology and the Pastoral Ideal in America*. New York: Oxford University Press, 1964. 392 pp. Index.

Deals with the response of American intellectuals to industrialism, especially the tension between the pastoral ideal of the simple, moral, self-sufficient life which was the core of the ideology of Jeffersonian

democracy, and the machine as the symbol of the mechanization of life. Finds that Americans developed a compromise
vision during the antebellum era in which the machine
became the emblem of progress, productivity, power over
nature, and the break with the past. The machine freed
the masses from want and gave Americans more time for
the life of the mind. Observes a continuing thread of
pessimism about the machine in the garden, however, in
the work of many nineteenth and twentieth-century American
writers.

127. Mayr, Otto. "Yankee Practice and Engineering Theory:
Charles T. Porter and the Dynamics of the
High-Speed Steam Engine." *Technology and
Culture*, 16 (1975): 570-602.

Documents the theoretical contributions of a
Yankee inventor whose public posture was complete innocence
of engineering and scientific theory. Finds that Porter
developed a dynamic theory of reciprocating engines
because he wanted to know why his empirically constructed
high-speed engine worked as well as it did. Shows that
Porter enjoyed the encouragement of trained engineers
and scientists.

128. Meier, Hugo A. "American Technology and the Nineteenth-
Century World." *American Quarterly*, 10
(1958): 116-130.

Observes that Americans saw their technological
achievements as a measure of the nation's progress and
a means of demonstrating the nation's values.

129. ————. "Technology and Democracy, 1800-1860."
Mississippi Valley Historical Review, 43
(1957): 618-640.

Argues that democracy and technology have had
a commensal symbiotic relationship in the United States.
Technology has provided the physical means of achieving
social, economic, and political equality, while the
American system has proven beneficial to the progress
of technology.

130. Morison, Elting E. *From Know-how to Nowhere: The Develop-
ment of American Technology*. New York: Basic
Books, 1974. xiii + 199 pp. Bibliography,
Index.

Suggests using historical data in decision-
making processes involving technological systems.

Presents the history of American technology post-1800 as
a microcosm of the dangers inherent in developing technological knowledge without the corresponding knowledge of
the proper utilization of technology. Argues that
twentieth-century technology has lost the close linkage
to human needs evident in earlier periods.

131. Oliver, John W. *History of American Technology*. New
             York: The Ronald Press, 1956. viii + 676 pp.
             Index.

Traces technology in the United States from the
European techniques imported by the first colonists to
the state of the art in the mid-twentieth century. Offers
little analysis. Contains numerous factual inaccuracies.

132. Ostrander, Gilman M. *American Civilization in the First
             Machine Age, 1890-1940*. New York and Evanston:
             Harper & Row, 1970. vii + 414 pp. Bibliography,
             Index.

Argues that the concurrence of the close of the
frontier with industrialization and mass immigration
transformed America from a patriarchy to a filiarchy--a
society that is continually developing technologically
and depends upon youth to acquire the necessary new
skills. Also views the United States as an ethnocracy--
a society of many racial, ethnic, and religious minorities,
divided along generational, rather than racial, ethnic,
or religious lines.

133. Pursell, Carroll W., Jr. *Early Stationary Steam Engines
             in America: A Study in the Migration of a
             Technology*. Washington, D.C.: Smithsonian
             Institution Press, 1969. viii + 152 pp.
             Bibliography, Index.

Contends that most histories of the American
Industrial Revolution--which claim that steam power was
unimportant in antebellum American industry--reflect a
bias towards events in New England, especially the
Massachusetts textile centers. Argues that the steam
engine was vital for manufacturing in locations where
waterpower was inadequate or unreliable, primarily the
South (sugar plantations) and the Old Northwest. Traces
the evolution of engine design and finds constant
communication across the Atlantic. Finds little difference between English and American designs, although the
United States constantly lagged behind English innovations
until the 1830s.

134. Rae, John B. "The 'Know-How' Tradition: Technology in American History." *Technology and Culture*, 1 (1960): 139-150.

Claims that American society was influenced to a unique degree by the forces of technology, in part because the economic and physical conditions of this country placed a high premium on accomplishing things quickly with the least amount of labor.

135. Roland, Alex. "Bushnell's Submarine: American Original or European Import?" *Technology and Culture*, 18 (1977): 157-174.

Rejects the traditional view of Bushnell as the originator of submarine warfare. Demonstrates that Bushnell adopted European ideas; he was not an inventor, but a transferrer of European technology.

136. Rosenberg, Nathan. "America's Rise to Woodworking Leadership." *America's Wooden Age: Aspects of its Early Technology*. Edited by Brooke Hindle. Tarrytown: Sleepy Hollow Restorations, 1975, pp. 37-62.

Argues that the technological innovations accompanying America's exploitation of its most valuable raw material were resource-intensive; i.e., the objective was the substitution of raw material for either labor or capital. Discusses the unique problems faced in utilizing woodworking machines in lieu of metalworking equipment.

137. ———. "Technological Interdependence in the American Economy." *Technology and Culture*, 20 (1979): 25-50.

Finds difficulties in quantitatively evaluating the role of technological innovation in the increase of productivity. Identifies intrinsic characteristics of technological change which are central to this difficulty: the need for complementary technology before a given invention is productive; the cumulative impact of small improvements; the difficulty in compartmentalizing the consequences of technological innovation within an industry.

138. ———. *Technology and American Economic Growth*.
              New York, Evanston, San Francisco, London:
              Harper & Row, 1972. xi + 211 pp. Index.

         Assembles an interpretive framework for under-
    standing how technology shaped the development of the
    American economy. Argues that the key factors in American
    technological development have been the high resource-
    labor ratio, producer initiative (producers accommodating
    their machines in the production of goods rather than
    accommodating the consumer, suppressing variation in
    design), and a high level of both technical and
    managerial skills, allowing for both innovation and
    the selective adoption of existing technology. Finds
    that the twentieth century has been characterized by the
    growing dependence of technology upon science.

139. Segal, Howard P. "American Visions of Technological
              Utopia, 1883-1933." *The Markham Review*, 7
              (1978): 65-76.

         Surveys twenty-five different visions of
    societies where the accelerated advance and spread of
    technology ultimately solved man's chronic material
    problems. These societies were not presented as impossible
    dreams but rather as the culmination of contemporary
    trends.

140. Sinclair, Bruce. *Philadelphia's Philosopher Mechanics:
              A History of the Franklin Institute, 1824-1865*.
              Baltimore and London: Johns Hopkins University
              Press, 1974. xi + 353 pp. Bibliography, Index.

         Examines the evolution of the Franklin Institute
    from its original ideal as a mechanics' institute which
    would encourage self-improvement through study and
    discussion, through its position as the preeminent
    technical center in the country circa 1840, to its role
    as arbitor for private industry in technical issues.
    Rejects the thesis that antebellum American technology
    was based essentially on the cut-and-try method. Finds
    that the Institute utilized experimental methods which
    united theory and practice.

141. Strassmann, W. Paul. *Risk and Technological Innovation:
              American Manufacturing Methods During the
              Nineteenth Century*. Ithaca: Cornell University
              Press, 1959. x + 249 pp. Bibliography, Index.

         Analyzes the interaction of business enterprise
    and technological change, concluding that American entre-

preneurs were creative but cautious innovators, rarely overestimating the chances for success. Presents four case studies: the iron and steel industry, the textile industry, the manufacturing of machine tools, and the development of electric power.

142. Temin, Peter. "Steam and Waterpower in the Early Nineteenth Century." *The Journal of Economic History*, 26 (1966): 187-205.

Concludes that steampower was not a dominant energy source during this period. Finds that it was widely utilized only in situations where it was cheaper to bring power to the raw material than to ship the raw material to waterpower sites.

143. Trachtenberg, Alan. **Brooklyn** *Bridge: Fact and Symbol*. New York: Oxford University Press, 1965. viii + 182 pp. Index.

Demonstrates how improvements in transportation have been key symbols in denoting American progress and the conquest of the wilderness. Views the Brooklyn Bridge as a cultural symbol of the transformation of the United States from a rural to an urban-industrial society. Analyzes J.A. Roebling's Hegalian leanings and their role in his vision of the Bridge as a monument and a symbol. Discusses the symbolic power of the Bridge in art and literature, concentrating on the poetry of Hart Crane.

144. Uselding, Paul J. "Henry Burden and the Question of Anglo-American Technological Transfer in the Nineteenth Century." *The Journal of Economic History*, 30 (1970): 312-337.

Analyzes the contributions of a mechanical engineer who was vital in introducing hot-working iron technology to the United States. Attacks the history of technology's preoccupation with invention rather than the diffusion of mechanical techniques.

145. Wallace, Anthony F.C. *Rockdale: The Growth of an American Village in the Early Industrial Revolution*. New York: Alfred A. Knopf, 1978. xx + 553 pp. Bibliography, Index.

Applies the methodology of cultural anthropology to a small, antebellum cotton-manufacturing town in Pennsylvania. Studies both the evolution of the technology and its social impact. Demonstrates that the

capitalists, drawing upon arguments from evangelical
Christianity, developed a form of religious paternalism
and a sense of moral stewardship which justified the
class structure, eased class conflict, and served as a
potent weapon against advocates of utopian socialism.

146. Welsh, Peter C.   "The Metallic Woodworking Plane: An
              American Contribution to Hand-Tool Design."
              Technology and Culture, 7 (1966): 38-47.

      Supports Kouwenhoven's thesis (see item 123)
by tracing the development of the plane from the wooden
English precedent, through minor modifications and hybrids,
to the simple, light, strong, and widely available
"vernacular expression of Victorian America" debuted
in 1876.

147. Wilkinson, Norman B.   "Brandywine Borrowings from European
              Technology." *Technology and Culture*, 4
              (1963): 1-13.

      Demonstrates how the textile, tanning, gunpowder,
and paper-making industries in a particular area were
dependent upon Europe for machinery and methodology
during the period prior to the first protective tariff
(1816).

148. York, Neil Longley.   "Technology in Revolutionary America,
              1760-1790." Ph.D. dissertation, University of
              California, Santa Barbara, 1978.

      Argues that the image of the Yankee tinker
obscures the fact that there was little technical knowledge
in colonial America. Finds little public understanding,
awareness, or appreciation of technology. Contends that
the war experience was responsible for a positive change
in public attitudes towards technology.

CHAPTER II: SPECIAL THEMES

EVOLUTION

149. Aldrich, Michele L. "United States: Bibliographic
        Essay." *The Comparative Reception of Darwinism*.
        Edited by Thomas F. Glick. Austin and London:
        University of Texas Press, 1975, pp. 207-226.

        Provides an evaluation of the literature.
    Finds that authors neglect the distinction between
    responses to Darwin's writings and responses to the
    forms of Darwinism developed by Darwin's supporters.

150. Allen, Garland E. "Thomas Hunt Morgan and the Problems
        of Natural Selection." *Journal of the
        History of Biology*, 1 (1968): 113-139.

        Discusses Morgan's initial rejection of
    Natural Selection because of the lack of a workable
    theory of heredity; he felt Darwin too speculative.

151. Bannister, Robert C. *Social Darwinism: Science and Myth
        in Anglo-American Social Theory*. Philadelphia:
        Temple University Press, 1979. ix + 292 pp.
        Index.

        Rejects the thesis that Darwin's ideas had
    been utilized by conservative leaders to justify the
    continued exploitation of the weak (see item 157).
    Contends that reformers labeled conservatives "social
    Darwinists" for polemic purposes. Sees the major
    influence of Darwinian concepts among the intervention-
    ists, who believed the needs of man transcended natural
    conditions.

152. Boromé, Joseph A.   "The Evolution Controversy." *Essays
            in American Historiography: Papers Presented
            in Honor of Allan Nevins*. Edited by Donald
            Sheehan and Harold C. Syrett. New York:
            Columbia University Press, 1960, pp. 169-192.

         Argues that the early reception of *Origin of
Species* was muted by the preoccupation with civil affairs
caused by the coming of the Civil War. Believes that it
was the generation of scientists reaching maturity after
the war who were more likely to embrace evolution. Notes
that there is little quantitative information about
American reading habits, placing limitations on discussions about the diffusion of ideas.

153. Bowler, Peter J.   "Edward Drinker Cope and the Changing
            Structure of Evolutionary Theory." *Isis*, 68
            (1977): 249-265.

         Discusses the shift in Cope's views on evolution
from a belief in a law of the acceleration of embryo-
logical growth to neo-Lamarckianism. Argues that this
shift enabled Cope to come to terms with the dominant
trends in late-nineteenth-century natural history.

154. Cravens, Hamilton.   "The Impact of Evolutionary Thought
            on American Culture in the 20th Century."
            Intellect, 106 (August 1977): 83-86.

         Claims that the impact of evolution has been
more pervasive in the twentieth century than in the
nineteenth, due to the efforts in the former century of
the social scientists, who have presented educated
Americans with a coherent evolutionary science of man
which promises social control.

155. Fleming, Donald.   "Social Darwinism." *Paths of American
            Thought*. Edited by Arthur M. Schlesinger, Jr.
            and Morton White. Boston: Houghton, Mifflin
            Company, 1963, pp. 123-146.

         Traces the evolution of social service careers,
such as the ministry, social work, and medicine, in the
wake of the redefinitions of social structures derived
from the writings of Herbert Spencer.

156. Gould, Stephen Jay.   "Agassiz's Later, Private Thought
              on Evolution: His Marginalia in Haeckel's
              Natürliche Schöpfungsgeschichte (1868)."   *Two
              Hundred Years of Geology in America: Proceed-
              ings of the New Hampshire Bicentennial Conference
              on the History of Geology* (item 461), pp.
              277-282.

        Presents these marginalia as Agassiz's most
coherent reaction to evolution during his later life.
Finds that Agassiz continued to argue that the geological
record empirically contradicted the theory of evolution,
and that the evidence for catastrophism still predominated.

157. Hofstadter, Richard. *Social Darwinism in American Thought*.
              Revised edition. Boston: Beacon Press, 1955.
              viii + 248 pp. Bibliography, Index.

        Analyzes the impact of Herbert Spencer's
writings on American thought and society. Finds that
many leading social theorists and businessmen utilized
Darwinian concepts to justify limiting government inter-
vention on behalf of the needy. Shows how similar
concepts were used to justify a racist and imperialistic
foreign policy at the end of the nineteenth century.
Concludes that Social Darwinism had fallen into disfavor
by the end of World War I because of a general reaction
against imperialistic and militaristic philosophies.

158. Loewenberg, Bert James.   "The Controversy over Evolution
              in New England, 1859-1873." *New England
              Quarterly*, 8 (1935): 232-257.

        Shows that during the years of intense
scientific debate over the validity of evolution,
orthodox Christians outside the scientific community
condemned the theory as part of an atheistic conspiracy.
Stresses that Darwin's theory appeared at a time when
orthodox religion was unsettled by critical Biblical
exegesis, studies in comparative religions, and changing
social conditions.

159. ————.   "Darwinism Comes to America, 1859-1900."
              Mississippi Valley Historical Review, 28
              (1941): 339-369.

        Stresses the coincidence of the arrival of
evolutionary theory and major social and economic
changes in the United States, resulting in intellectual
and social atmospheres unfavorable to absolutes.

160. ———. "The Reaction of American Scientists to
            Darwinism" *American Historical Review*, 38
            (1933): 687-701.

    Focuses on three representative figures in
American science. Louis Agassiz opposed Darwin's theory
on scientific grounds. Supporters of the theory were
led by Asa Gray. James Dwight Dana was an example of a
scientist who initially rejected evolution but eventually
changed his position.

161. Mayr, Ernst. "Agassiz, Darwin, and Evolution." *Harvard
            Library Bulletin*, 13 (1959): 165-194.

    Argues that Agassiz's education and intellectual
background made him unsympathetic to Darwin's theory of
evolution. Identifies four dominant characteristics of
Agassiz's thinking: belief in a rational plan for the
universe, a typological approach to species, an expecta-
tion that the universe was discontinuous, and an onto-
genetic conception of evolution. This argument is
critiqued in item 174.

162. Moore, James R. *The Post-Darwinian Controversies: A
            Study of the Protestant Struggle to Come to
            Terms with Darwin in Great Britain and America,
            1870-1900*. Cambridge, London, New York,
            Melbourne: Cambridge University Press, 1979.
            xi + 502 pp. Bibliography, Index.

    Rejects the military metaphors previously used
to describe the relationship between post-Darwin science
and religion, offering in their place the nonviolent
perspective that science and religion are congenial.
Analyzes the thought of twenty-eight Christian controver-
salists and concludes that orthodox Darwinism was
accepted only by those who believed in an orthodox
theology; liberal theologians embraced other forms of
evolution, while the utter rejection of evolution was
based generally on scientific and philosophical grounds
rather than on appeals to Scripture. Includes a very
extensive and useful bibliography.

163. Moore, John A. "Creationism in California." *Daedalus*,
            103, No. 3 (Summer 1974): 173-189.

    Traces the conflict between creationists and
advocates of evolution over the use of the Biological
Sciences Curriculum Study high school textbooks.
California was studied because it purchases ten percent

of all public school texts and has a large number of
vocal creationists; hence, it is a bellweather state.

164. Numbers, Ronald L. *Creation by Natural Law: Laplace's
Nebular Hypothesis in American Thought.* Seattle
and London: University of Washington Press,
1977. xi + 184 pp. Bibliography, Index,
Appendices.

Traces the history of the hypothesis from its
creation as a cosmogony free from Scriptural influence,
through its introduction to the American reading public
in the Bridgewater Treatises and adoption by the American
scientific community in the wake of Daniel Kirkwood's
formulation of his analogy, to the debates over its
validity in the late nineteenth century. Finds that
part of the appeal of the hypothesis was the ease in
which it could be reconciled with natural theology.
Argues that Darwin's theory was rapidly assimilated by
Americans because the nebular hypothesis had already
established the precedent that a theory of natural
development need not represent a threat to Christianity.

165. ————. "Science Falsely So-Called: Evolution and
Adventists in the Nineteenth Century."
Journal of the American Scientific Affiliation,
27 (1975): 18-23.

Finds that the Seventh-Day Adventists rejected
evolutionary theories because of conflicts with their
literal reading of Scripture. Examines the extreme
Baconian vision of science adopted by the Adventists.

166. Persons, Stow, editor. *Evolutionary Thought in America.*
New Haven: Yale University Press, 1950.
Reprint. Hamden, Conn.: Archon Books, 1968.
x + 462 pp. Index.

Discusses the impact of evolutionary thought
on sociology, politics, economics, psychology, literature,
architecture, moral theory, and theology.

167. Pfeifer, Edward J. "The Genesis of American Neo-
Lamarckism." *Isis,* 56 (1965): 156-167.

Describes neo-Lamarckism as "an attack upon
natural selection as the primary factor in evolution."
Traces the theory back to its independent formulations
by Alpheus Hyatt and Edward Drinker Cope. Claims that
Louis Agassiz, not Lamarck, was the major influence upon
the thinking of the neo-Lamarckians.

168. ————. "United States." *The Comparative Reception of
        Darwinism*. Edited by Thomas F. Glick. Austin
        and London: University of Texas Press, 1974,
        pp. 168-206.

     Discusses the fates of evolutionary theories in
the United States from Chambers forward. Observes that
the acceptance of evolution by the mid-1870s was not a
victory for natural selection, but rather, for neo-
Lamarckianism, which was viewed as an American theory,
appealed to the religious-minded, and could be pressed
into the service of reform social activities as a counter
to the laissez faire attitude of Social Darwinists.

169. Provine, William B. "Francis B. Sumner and the Evolution-
        ary Synthesis." *Studies in History of Biology*,
        3 (1979): 211-240.

     Claims that Sumner was the first researcher to
combine the techniques of the naturalist-systematist and
the laboratory geneticist. His studies of the geographical
races of the deer mouse, utilizing experimental individuals
from natural populations, led to a consensus among the
two groups of scientists regarding the mechanism of
evolution.

170. Russett, Cynthia Eagle. *Darwin in America: The Intellectual
        Response, 1865-1912*. San Francisco: W.H.
        Freeman and Co., 1976. ix + 228 pp. Index.

     Utilizes printed sources to provide a survey
of the impact of evolution upon religious thought,
philosophy, social thought, history, economics, and
literature. In some cases the focus is very narrow;
Henry Adams is the sole representative of history, while
Thorstein Veblen speaks for economics. Fails to provide
a synthetic evaluation of that impact or demonstrate an
understanding of the scientific ideas which were
impacting upon the intelligentsia.

171. Sanford, William F., Jr. "Dana and Darwinism." *Journal
        of the History of Ideas*, 26 (1965): 531-546.

     Examines James Dwight Dana's transformation
from a supporter of the concept of the immutability of
species to a believer in natural selection. Identifies
his insistence on the necessity of God's special inter-
vention in the creation of man as his major continuing
difference with Darwin.

172. Stocking, George W., Jr. "Lamarckianism in American
        Social Science, 1890-1915." *Journal of the
        History of Ideas*, 23 (1962): 239-256.

    Identifies the Lamarckian attributes of the
theories of a number of American social scientists.
Credits the rejection of Lamarckianism by biologists
with destroying the last important theoretical link
between social and biological theory, leaving the social
sciences to develop independently of, rather than sub-
ordinately to, biology.

173. Wiener, Philip P. *Evolution and the Founders of
        Pragmatism*. Cambridge, Mass.: Harvard
        University Press, 1949. xiv + 288 pp.
        Appendices, Index.

    Identifies the founding fathers of pragmatism
as members of an informal gathering of social and natural
scientists known as The Metaphysical Club. Contends that
the combined impact of Darwinian evolution and the rise
of statistical concepts in physics led these men to view
nature as contingent, and truth as probable, relative,
temporal, and fallible.

174. Winsor, Mary Pickard. "Louis Agassiz and the Species
        Question." *Studies in History of Biology*,
        3 (1979): 89-117.

    Disputes the claim in item 161 that Agassiz
rejected evolution on philosophical grounds. Contends
that he was too dogmatic to admit error and revise his
position once he had declared the fixity of species.

FUNDING

175. Burstyn, Harold L. "Seafaring and the Emergence of
        American Science." *The Atlantic World of
        Robert G. Albion*. Edited by Benjamin W.
        Labaree. Middletown: Wesleyan University
        Press, 1975, pp. 76-109, 223-228.

    Treats American ocean shipping as the economic
base upon which antebellum scientific institutions
developed.

176. Coben, Stanley. "American Foundations as Patrons of
         Science: The Commitment to Individual Research."
         *The Sciences in the American Context: New
         Perspectives* (item 84), pp. 229-247.

         Describes the post-World War I decision by
     private foundations to support individual scientists "one
     of the most significant events in the history of learning
     in the United States." Views Wickliffe Rose (Rockefeller
     Foundation), Beardsley Ruml (Laura Spelman Memorial Fund),
     and Henry Allen Moe (Guggenheim) as the chief architects
     of this innovation.

177. Davis, Lance E., and Daniel J. Kevles. "The National
         Research Fund: A Case Study in the Industrial
         Support of Academic Science." *Minerva*, 12
         (1974): 207-220.

         Contends that most industrial leaders refused
     to contribute to this fund because everybody would have
     benefited from the research, whether or not they contri-
     buted. Finds that only firms in monopoly or near-
     monopoly positions, as well as trade associations, could
     afford such an investment.

178. Greenberg, Daniel S. *The Politics of Pure Science*.
         New York: The New American Library, 1967.
         xiii + 303 pp. Index.

         Focuses on the period after World War II.
     Identifies a scientific elite but denies the existence
     of a scientific establishment. Finds two ideologies
     bonding the scientific community: the desire for society
     to support but not govern research and the desire for
     the scientific community to exist in a state of
     "meritocratic anarchy." Studies the Mohole fiasco
     and the debates over high-energy particle accelerators
     to demonstrate that the demands by Congress for account-
     ability increased in proportion to the size of the
     public contributions to scientific budgets.

179. Kevles, Daniel J. "Millikan: Spokesman for Science in
         the Twenties." *Engineering and Science*, 32
         (1969): 17-22.

         Describes Millikan's strategy for encouraging
     business leaders to support basic research. Finds that
     he presented basic research, which, he argued, eventually
     leads to greater material abundance, as a superior
     alternative to the redistribution of wealth in meeting

the needs of the poor. Contends that Millikan's close
links to the business community and identification with
conservative policies led to his condemnation during
the Great Depression.

180. Kohler, Robert E., Jr. "Warren Weaver and the Rockefeller
Foundation Program in Molecular Biology: A
Case Study in the Management of Science."
*The Sciences in the American Context: New
Perspectives* (item 84), pp. 249-293.

Argues that Weaver, director of the natural
science division of Rockefeller, created the role of
manager of science--i.e., an individual concerned with
the formulation of policy and overview of activity.
Finds that he did not create a new institutional setting,
but rather, established a new relationship between patron
and academic science, overseeing academic research from
outside academia. Emphasizes that Weaver encouraged the
application of the physical sciences to biological problems.

181. Miller, Howard S. *Dollars for Research: Science and Its
Patrons in Nineteenth-Century America*. Seattle
and London: University of Washington Press,
1970. xiv + 258 pp. Bibliography, Index.

Explores the many ways in which the private
sector funded scientific activities from the establishment
of the Smithsonian Institution to the founding of the
Carnegie Institution of Washington. Argues that private
support, although quantitatively less than the funds
received from public sources, was more important qualitatively because private donors were more flexible, more
likely to aid very abstract research, and more willing
to support innovation. Concludes that American science
was successful in attracting private support and rejects
the indifference argument.

182. ————. "The Political Economy of Science." *Nineteenth-
Century American Science: A Reappraisal* (item
32), pp. 95-112.

Denies that the scientific community was
married to a laissez faire philosophy towards governmental
funding of science. Finds that scientists were willing
to, and in practice did, accept financial support from
a mixture of public and private sources.

183. ———. "Science and Private Agencies." *Science and
            Society in the United States* (item 99),
            pp. 191-221.

     Examines the role of private philanthropy in
shaping scientific research in the United States. Emphasizes the activities of the Smithsonian and the Carnegie
Institution of Washington.

184. Plotkin, Howard. "Edward C. Pickering and the Endowment
            of Scientific Research in America, 1877-1918."
            Isis, 69 (1978): 44-57.

     Argues that Pickering was a transitional figure
in the history of patronage of American science: he
recognized the expanding needs of the American scientific
community in the twentieth century but was unable to
devise a successful mechanism. Blames his failure on
his inability to recognize the inappropriateness for
other sciences of the methods used to raise funds for
astronomy.

185. Reingold, Nathan. "The Case of the Disappearing
            Laboratory." *American Quarterly*, 29 (1977):
            79-101.

     Corrects the standard account of the establish-
ment of the National Research Council Fellowships.
Discusses the rejection of a central research laboratory
for physics and chemistry in favor of the fellowships.

186. ———. "National Science Policy in a Private Foundation:
            The Carnegie Institution of Washington." *The
            Organization of Knowledge in Modern America,
            1860-1920*. Edited by Alexandra Oleson and
            John Voss. Baltimore and London: The Johns
            Hopkins University Press, 1979, pp. 313-341.

     Argues that the lack of alternative basic
research centers in the nation resulted in the Carnegie
dealing with national policy issues, especially the
question of establishing research priorities. Finds
that the Carnegie took a cautious attitude towards the
priority question, selecting fields which were already
well established in the United States and contained an
exceptional researcher.

187. Struik, D.J. "The Influence of Mercantilism on Colonial
        Science in America." *Organon*, 1 (1964): 157-
        163.

    Argues that the pattern of scientific activity
in the colonies reflected the needs of the dominant
economic force, the merchants. Includes astronomy,
mathematics, optics, magnetism, and descriptive natural
history (Newtonian and Linnaean sciences) under the
rubric of mercantilist sciences.

188. Tobey, Ronald C. *The American Ideology of National
        Science, 1919-1930*. Pittsburgh: University of
        Pittsburgh Press, 1971. xiii + 263 pp.
        Bibliography, Index.

    Focuses on an attempt by the scientific
community to impress upon the public the relevance of
professional science, especially pure research, to
broader cultural and political values--such as progressive
democracy--in hope of developing lay sympathy for science.
Examines the activities of Science Service, a news
service which both highlighted the impact of scientific
ideas upon everyday life and culture and popularized
scientific discoveries. Concludes that the attempt
failed because of public attitudes towards Einstein's
Theory of Relativity; Einstein's concepts violated common
sense and once again proved to the public that science
was a field only decipherable by the expert. Views the
collapse of the campaign for the National Research
Endowment as the final indication that the scientific
community failed to link pure research with larger
national ideologies.

LEARNED SOCIETIES

189. Bates, Ralph S. *Scientific Societies in the United
        States*. 3d edition. Cambridge, Mass.: The
        M.I.T. Press, 1965. iv + 326 pp. Chronology,
        Bibliography, Index.

    Surveys the development of scientific and
technical societies and their role in the coordination
and stimulation of scientific activity. Finds that
scientific societies reflected the changing needs and
conditions of the scientific community. Examines the
multitude of local and regional societies active during
the antebellum period. Identifies three trends during
the years 1866-1918: specialization, centralization, and

the rise of technological societies. Observes increasing
international involvement in the years since World War I.

190. Bell, Whitfield J., Jr. "The American Philosophical
            Society as a National Academy of Sciences,
            1780-1846." *Proceedings of the Tenth Inter-*
            national Congress of the History of Science.
            Paris: Hermann, 1964, 1:165-177.

        Credits the APS with filling the void created
by the reluctance of the federal government to create
national scientific institutions. Finds that the APS
served as the national library, museum, and academy of
science. Claims it lost its preeminence in part because
it remained a catholic intellectual center at a time of
increasing specialization, in part because Philadelphia
had lost its position as the nation's first city, and
in part because the government finally established its
own network of scientists and scientific institutions.

191. ————. "The Cabinet of the American Philosophical
            Society." *A Cabinet of Curiosities: Five*
            Episodes in the Evolution of American Museums
            (item 103), pp. 1-34.

        Observes that the APS spent the first century
of its existence acquiring objects without any collecting
rationale beyond the belief that a learned society had
the obligation to maintain a museum; the result was a
disorganized clutter. The second century (post-1849)
was spent dispersing the holdings among more specialized
museums.

192. Cochrane, Rexmond Canning. *The National Academy of*
            Sciences: The First Hundred Years, 1863-1963.
            Washington, D.C.: National Academy of Sciences,
            1978. xv + 694 pp. Appendices, Index.

        Presents an official history based upon the
Academy's archives. Concentrates on the relationship
between the Academy and the federal government, as the
Academy slowly expanded its advisory role.

193. Dupree, A. Hunter. "The National Academy of Sciences and
            the American Definition of Science." *The*
            Organization of Science in Modern America,
            1860-1920. Edited by Alexandra Oleson and
            John Voss. Baltimore and London: The Johns
            Hopkins University Press, 1979, pp. 342-363.

Views the Academy as a rallying point for American science and repository of values. Contends it functioned as a social unit, a mobilizer of power and resources, and an exemplar.

194. ————. "The National Pattern of American Learned Societies." *The Pursuit of Knowledge in the Early American Republic: American Scientific and Learned Societies from Colonial Times to the Civil War* (item 209), pp. 21-32.

Defines learned societies as information systems whose responsibilities are the gathering, processing, and dissemination of knowledge. Demonstrates how American learned societies fit that definition.

195. Ewan, Joseph. "The Growth of Learned and Scientific Societies in the Southeastern United States to 1860." *The Pursuit of Knowledge in the Early American Republic: American Scientific and Learned Societies from Colonial Times to the Civil War* (item 209), pp. 208-218.

Describes these societies as friendship leagues formed to encourage communication. Views such societies as essentially urban phenomena and rare in the sparsely populated South.

196. Flack, J. Kirkpatrick. *Desideratum in Washington: The Intellectual Community in the Capital City, 1870-1900.* Cambridge, Mass.: Schenkman Publishing Company, 1975. x + 192 pp. Bibliography, Index.

Attempts to analyze the rise of local scientific organizations in Washington, such as the Philosophical Club of Washington and the Washington Academy of Sciences, within the context of an expanding intellectual community and a desire to institutionalize intellectual life. Provides useful histories of these local organizations but is less successful in its discussion of more general themes.

197. Frick, George F. "The Royal Society in America." *The Pursuit of Knowledge in the Early American Republic: American Scientific and Learned Societies from Colonial Times to the Civil War* (item 209), pp. 70-83.

Examines the activities of the Royal Society of London, which served as a learned society for the British colonies well into the eighteenth century.

198. Gerstner, Patsy A. "The Academy of Natural Sciences of Philadelphia, 1812-1850." *The Pursuit of Knowledge in the Early American Republic: American Scientific and Learned Societies from Colonial Times to the Civil War* (item 209), pp. 174-193.

Divides the early history of the Academy into three stages: the amateur collecting period, 1812-1817; the Maclure era, 1817-1840, under the paternalistic leadership of William Maclure; and the transitional phase after Maclure's death in 1840 when the Academy began to evolve into a professional scientific organization.

199. Greene, John C. "Science, Learning, and Utility: Patterns of Organization in the Early American Republic." *The Pursuit of Knowledge in the Early American Republic: American Scientific and Learned Societies from Colonial Times to the Civil War* (item 209), pp. 1-20.

Finds that learned societies tended to be local or regional, with overlapping memberships in organizations within the same region preventing competition and encouraging mutual support. Argues that Philadelphia organizations served as the model for other cities.

200. Gross, Walter E. "The American Philosophical Society and the Growth of Science in the United States, 1835-1850." Ph.D. dissertation, University of Pennsylvania, 1970.

Argues that the APS remained important for American science only as long as there were no alternative organizations; when alternatives arose in the 1840s, the APS rapidly became insignificant. Finds that the APS lacked the money, equipment, and organizational spirit to coordinate research. Provides accounts of the scientific work of members of the APS.

201. Harris, Jonathan. "New York's First Scientific Body: The Literary and Philosophical Society, 1814-1834." *Annals of the New York Academy of Sciences*, 196 (1972): 327-337.

> Blames the demise of the society on its honorific nature and the death or withdrawal of the founding leaders. Finds that it was eclipsed by the Lyceum of Natural History.

202. Hendrickson, Walter B. "Science and Culture in the American Middle West." *Isis*, 64 (1973): 326-340.

> Surveys thirteen antebellum Midwestern academies of science and concludes that their members were usually physicians or amateur naturalists; these academies typically supported publication programs, held regular meetings, and attempted to establish museums. Argues that these academies were part of the attempt to reproduce Eastern urban culture in the Middle West and reflected general trends in American science.

203. Hindle, Brooke. "The Underside of the Learned Society in New York, 1754-1854." *The Pursuit of Science in the Early American Republic: American Scientific and Learned Societies from Colonial Times to the Civil War* (item 209), pp. 84-116.

> Attributes New York City's failure to establish a general scientific society to its repeated decision to model its learned societies upon the London Society of Arts, which was concerned with commercial products, rather than the Royal Society of London.

204. Kerber, Linda K. "Science in the Early Republic: The Society for the Study of Natural Philosophy." *William and Mary Quarterly*, 3d ser. 29 (1972): 263-280.

> Examines one example of a fairly common phenomenon at the beginning of the nineteenth century—young professionals forming a club for the purposes of self-education and the diffusion of scientific knowledge. Claims that these clubs succeeded in sensitizing their members to the needs of science, despite their typically short lives.

205. Kohlstedt, Sally Gregory. "From Learned Society to
            Public Museum: The Boston Society of Natural
            History." *The Organization of Knowledge in
            Modern America, 1860-1920*. Edited by Alexandra
            Oleson and John Voss. Baltimore and London:
            The Johns Hopkins University Press, 1979,
            pp. 386-406.

        Views the society as a relic of an older age
    which had lost its relevance for the productive scientific
    community and, as an alternative, pioneered the develop-
    ment of a scientific museum for the general public.

206. ———. "The Nineteenth-Century Amateur Tradition: The
            Case of the Boston Society of Natural History."
            Science and Its Public. Edited by Gerald
            Holton and William A. Blanpied. Dordrecht:
            D. Reidel, 1976, pp. 173-190.

        Studies the Boston Society during its transfor-
    mation from a voluntary association of active amateur
    naturalists to a center for the education of the general
    public.

207. ———. "A Step Towards Scientific Self-Identity in the
            United States: The Failure of the National
            Institute, 1844." *Isis*, 62 (1971): 339-362.

        Views the inability of the National Institute
    to garner widespread support among scientists for its
    1844 meeting as an indication that its program for
    organizing American science (open membership and strong
    political ties) was doomed.

208. Nodyne, Kenneth R. "The Founding of the Lyceum of Natural
            History." *Annals of the New York Academy of
            Sciences*, 172 (1970): 139-149.

        Argues that the Lyceum was the end-product of
    cultural rivalry between New York and Philadelphia,
    personal rivalry between Samuel L. Mitchill and DeWitt
    Clinton, and the intellectual rivalry between naturalists
    and the universal scholars. Stresses that it was
    established at a time of unprecedented cultural national-
    ism and economic growth for New York.

209. Oleson, Alexandra, and Sanborn C. Brown, editors. *The Pursuit of Knowledge in the Early American Republic: American Scientific and Learned Societies from Colonial Times to the Civil War.* Baltimore and London: The Johns Hopkins University Press, 1976. xxv + 372 pp. Index.

Contains eighteen articles delineating the history of learned societies in the United States and Canada. Includes items 194, 195, 197, 198, 199, 203, 211, 241, 736.

210. Pauly, Philip J. "The World and All That Is in It: The National Geographic Society, 1888-1918." *American Quarterly*, 31 (1979): 517-532.

Examines the evolution of the society from a professional scientific organization to a scientific society run by and for the nonspecialist interested in natural history. Finds that the society was shaped by the genteel, catholic vision of science held by Alexander Graham Bell, the chief benefactor and president of the society during the transition, and Gilbert Grosvenor, editor of the society's journal.

211. Shapiro, Henry D. "The Western Academy of Natural Sciences of Cincinnati and the Structure of Science in the Ohio Valley, 1810-1850." *The Pursuit of Knowledge in the Early American Republic: American Scientific and Learned Societies from Colonial Times to the Civil War* (item 209), pp. 219-247.

Argues that this academy faded away after 1850 because the changing scientific community had developed new needs which it could not fill. It was oriented to support local, amateur taxonomic investigations at a time of increasing professionalization and nationalization of science.

212. Shipton, Clifford K. "The Museum of the American Antiquarian Society." *A Cabinet of Curiosities: Five Episodes in the Evolution of American Museums* (item 103), pp. 35-48.

Concludes that the failure to establish a firm collecting policy led to overcrowding and lack of organization. Argues that such conditions greatly reduced the value and utility of the society's holdings.

POLLUTION, ECOLOGY, AND CONSERVATION

213. Burgess, Robert L. "The Ecological Society of America:
               Historical Data and Some Primary Analyses."
               History of American Ecology (item 219), separate
               pagination, 24 pp.

          Presents an overview of membership growth;
     geographical distribution of members, presidents, and
     meetings; publications; awards; and committees. Finds
     fairly steady growth from the founding of the society in
     1914 through 1950, then an acceleration due to the general
     growth of American science and expanding employment
     opportunities for ecologists. Views the leadership and
     the membership as dominated by the eastern third of the
     nation.

214. Burke, John G. "Wood Pulp, Water Pollution, and Adver-
               tising." *Technology and Culture*, 20 (1979):
               175-195.

          Links the increase in water pollution in the
     United States between 1870 and 1920 with the rise in
     mass-market advertising and its use of immense quantities
     of wood pulp. Shows that the wood pulp was produced by
     a process which generated large quantities at low cost,
     but with the by-product of massive water pollution.

215. Cain, Louis P. "Unfouling the Public's Nest: Chicago's
               Sanitary Diversion of Lake Michigan Water."
               Technology and Culture, 15 (1974): 594-613.

          Examines attempts to control sewage pollution
     in Chicago from 1889 to 1930, focusing on policy issues,
     the interaction of government units, and technological
     and economic constraints. Describes the change in
     approach from sewage disposal to sewage treatment.

216. Cittadino, Eugene. "Ecology and the Professionalization
               of Botany in America, 1890-1905." *Studies in
               History of Biology*, 4 (1980): 171-198.

          Approaches plant ecology as a dominant specialty
     in botany and a consequence of the maturation and pro-
     fessionalization of American botany; it was an extension
     of plant physiology to the plant in its natural setting.
     Rejects the theses that ecology was either a response to
     the needs of agriculture or a recognition of the limita-
     tions of natural resources.

217. Dodds, Gordon B. "The Stream-Flow Controversy: A
              Conservation Turning Point." *Journal of
              American History*, 56 (1969): 56-69.

       Studies the attack issued by the Corps of
Engineers against the conservationists' theory that
deforestation radically affects stream flow, leading to
erosion and floods. Finds that the Corps utilized
scientific observations and methods, while the conservationists relied on emotionalism and dogmatism. Suggests
that historians have been too generous in their praise
of conservationists and not appreciative enough of the
positions of their opponents.

218. Egerton, Frank N. "Ecological Studies and Observations
              Before 1900." *Issues and Ideas in America*.
              Edited by Benjamin J. Taylor and Thurman J.
              White. Norman: University of Oklahoma Press,
              1976, pp. 311-351.

       Surveys the history of ecology prior to its
establishment as a formal discipline. Contends that
American researchers were generally behind the Europeans.
Identifies studies of the influence of the environment
on health as the first research program in ecology.

219. ———, editor. *History of American Ecology*. New York:
              Arno Press, 1977. iii + 270 pp.

       Reprints seven essays and offers two previously
unpublished essays. Includes items 213, 218, 222, 224,
225, 227, 618.

220. Emmons, David M. "Theories of Increased Rainfall and
              the Timber Culture Act of 1873." *Forest
              History*, 15 (1971): 6-14.

       Describes the scientific support for the theory
that forestation could lead to a modification of the
climate, the basis for the legislation which granted
land in exchange for the cultivation of trees on the
Great Plains.

221. Flader, Susan L. *Thinking Like a Mountain: Aldo Leopold
              and the Evolution of an Ecological Attitude
              Toward Deer, Wolves, and Forests*. Columbia:
              University of Missouri Press, 1974. xxv +
              284 pp. Bibliography, Index.

       Focuses on Leopold's changing attitude towards
methods of preserving the wilderness ecosystem, which

culminated in his 1944 essay "Thinking Like a Mountain."
Assesses the role of observation, experience, science,
philosophy, policy, and politics in his development of
an ecological attitude. Contends that the key issue
facing him as a Conservation Commissioner was the substitution of hunting for the efforts of natural predators
in checking the deer population; this was a politically
unpopular but scientifically defendable position.

222. Frey, David G. "Wisconsin: The Birge-Juday Era."
     Limnology in North America. Madison: University
     of Wisconsin Press, 1963, pp. 3-54.

Examines the work of the two men--E.A. Birge and
Chancey Juday--who dominated descriptive limnology from
1875 to 1941. Traces the evolution of their work from
the study of the systematics and distribution of fresh-
water plankton, through explorations of the physical and
chemical causes for the distribution, to the conceptuali-
zation of a lake as an organism which should be studied
as an ecological unit.

223. Hays, Samuel P. *Conservation and the Gospel of Efficiency:
     The Progressive Conservation Movement, 1890-
     1920*. Cambridge, Mass.: Harvard University
     Press, 1959. xii + 297 pp. Bibliography,
     Index.

Views the conservation movement as an attempt
to apply scientific knowledge to practical problems,
reflecting optimism, confidence, and a felt need for
rational planning. Finds the roots of conservation in
attempts to restrict resource exploitation in the West.
Examines the expansion of these efforts into a moral
crusade for efficient use of all natural resources to
guarantee sufficent reserves for future generations.
Contends that the movement was not anti-monopoly; its
concern was rational usage, not the question of private
versus public ownership.

224. McIntosh, Robert P. "Ecology Since 1900." *Issues and
     Ideas in America*. Edited by Benjamin J.
     Taylor and Thurman J. White. Norman:
     University of Oklahoma Press, 1976, pp. 353-
     372.

Traces the evolution of the science in America
through the rise of the New Ecology in the 1950s, pointing
out benchmark conferences and publications.

225. ———. "H.A. Gleason--'Individualistic Ecologist' 1882-
          1975: His Contributions to Ecological Theory."
          Bulletin of the Torrey Bontanical Club, 102
          (1975): 253-273.

          Deals with Gleason's revolutionary concept that
     no two communities were identical because no two species
     make the same demands upon, and hence create the same
     niche in, an environment; an association of flora was
     not a supra-organism, as most ecologists had argued, but
     rather, a coincidence.

226. Nelkin, Dorothy. "Science and Professional Responsibility:
          The Experience of American Ecologists." *Social
          Studies of Science*, 7 (1977): 75-95.

          Distinguishes between scientific organizations,
     whose traditional responsibility has been the maintenance
     of standards within the discipline, and professional
     guilds, which license, regulate, and control practitioners,
     as well as oversee practitioner-client relationships.
     Finds that ecologists are trying to find a middle ground
     between scientific organizations and professional guilds
     as public demands for their expertise raise questions
     about their social responsibility.

227. Tobey, Ronald. "American Grassland Ecology, 1895-1955:
          The Life Cycle of a Professional Research
          Community." *History of American Ecology* (item
          219), separate pagination, 56 pp.

          Tests successfully the Crane-Price hypothesis
     of the shape of the growth curve of the literature of a
     scientific field and the Crane-Kuhn hypothesis of the
     developmental stages of the social structure of a
     scientific field.

228. ———. "Theoretical Science and Technology in American
          Ecology." *Technology and Culture*, 17 (1976):
          718-728.

          Denies that ecology was an outgrowth of European
     studies of plant geography. Asserts that it was the
     result of a native shift from a static to a dynamic
     perspective of the plant. This new view derived from
     technological efforts to control vegetational changes
     on the Midwest grasslands.

229. Wadland, John Henry. *Ernest Thompson Seton: Man in
         Nature and the Progressive Era, 1880-1915.*
         New York: Arno Press, 1978. xii + 528 pp.
         Bibliography.

     Evaluates Seton's contributions, through nature
stories and animal life histories, to the development of
an ecology and an ethology which rejected the anthropo-
centric view of life. Defends the accuracy and scientific
competence of Seton's work. Does not attempt to be a
definitive biography and ignores large parts of Seton's
personal life.

POPULAR SCIENCE AND POPULARIZATION

230. Carter, Paul A. "Science and the Common Man." *The
         American Scholar.* 45 (1975-76): 778-794.

     Surveys the attitude of non-scientists towards
science and of scientists towards popular understanding
of science during the 1920s. Finds public attitudes
ranging from opposition to research to hero worship of
Einstein; scientific attitudes ranged from a belief that
science was unfathomable to the public to conscious
efforts at popularization.

231. Greene, John C. "Science and the Public in the Age of
         Jefferson." *Isis*, 49 (1958): 13-25.

     Argues that during the late eighteenth and
early nineteenth centuries, an age in which popular
science was generally not held in high esteem, the
situation in the United States was even less favorable.
Finds little demand for popular lectures and little
desire on the part of the American scientific community
to present them.

232. Overfield, Richard A. "Science in the *Virginia Gazette*,
         1736-1780." *The Emporia State Research Series*,
         16, No. 3 (1968).

     Finds the coverage of scientific events pro-
vided by a typical colonial newspaper somewhat uneven.
There were detailed accounts of astronomy (e.g., the
Transit of Venus) and electricity (Franklin's work), but
little discussion of natural history and chemistry.

233. Rossiter, Margaret W. "Benjamin Silliman and the Lowell
        Institute: The Popularization of Science in
        Nineteenth-Century America." *The New England
        Quarterly*, 44 (1971): 602-626.

    Evaluates the Lowell Institute and argues that
the Lowell Lectures reveal the level of America's support
for quality popular lectures on scientific topics. Finds
that the content of Silliman's lectures compares favorably
to those offered in college courses.

234. Stowell, Marion Barber. *Early American Almanacs: The
        Colonial Weekday Bible*. New York: Burt Franklin,
        1977. xviii + 331 pp. Bibliography, Appen-
        dices, Index.

    Emphasizes the dominant role of the almanac as
a source of secular information in colonial America.
Finds that the scientific and academic philomath almanacs
of the seventeenth century were replaced in the eighteenth
century by farmer's almanacs. The latter added astrology,
humor, satire, and practical advice for farmers while
offering less learned essays on scientific topics.
Philomath almanacs described the Copernican conception
of the universe, while the farmer's almanacs popularized
Newtonian science and the image of God as Clockmaker.

235. Tucker, Louis Leonard. "'Ohio Show-Shop': The Western
        Museum of Cincinnati, 1820-1867." *A Cabinet
        of Curiosities: Five Episodes in the Evolution
        of American Museums* (item 103), pp. 73-105.

    Traces the decline of an institution founded to
provide popular science education into an entertainment
center with a science facade.

236. Whalen, Matthew, and Mary F. Tobin. "Periodicals and the
        Popularization of Science in America, 1860-
        1910." *Journal of American Culture*, 3
        (1980): 195-203.

    Contends that periodicals were the dominant
source of popular scientific information during this
period. Divides these periodicals into three types:
general science, scientific study, and popular science.
Discusses their respective audiences using an expanded
version of Reingold's classification scheme of the
scientific community (see item 241). The authors have
misunderstood Reingold, leading to some confusing
conclusions.

237. Zochert, Donald. "The Natural Science of an American
                 Pioneer: A Case Study." *Transactions of the
                 Wisconsin Academy of Sciences, Arts and Letters*,
                 60 (1972): 7-15.

        Finds that a considerable amount of elementary
     science, especially descriptive natural history, was
     known by a typical literate but uneducated frontiersman.

238. ———   "Science and the Common Man in Ante-Bellum
                 America." *Isis*, 65 (1974): 448-473.

        Analyzes newspaper coverage of science between
     1837 and 1846. Concludes that the reader was exposed
     generally, but not exclusively, to the exotic and novel
     aspects of science. Astronomy was the most thoroughly
     treated discipline. Finds that science was supported
     for its social utility--the power to upgrade the cultural
     and moral level of an individual or community.

PROFESSIONALIZATION

239. Daniels, George H. "The Process of Professionalization
                 in American Science: The Emergent Period,
                 1820-1860." *Isis*, 58 (1967): 151-166.

        Identifies four stages in the professionalization
     process--preemption, institutionalization, legitimation,
     and the attainment of professional autonomy--and contends
     that the first three stages form the emergent period.

240. Holmfeld, John D. "From Amateurs to Professionals in
                 American Science: The Controversy over the
                 Proceedings of an 1853 Scientific Meeting."
                 *Proceedings of the American Philosophical
                 Society*, 114 (1970): 22-36.

        Views the supression of the original proceedings
     of the Cleveland meeting of the American Association for
     the Advancement of Science by the AAAS's leadership as
     an effort by the professional scientific community to
     forestall possible European ridicule of the highly
     speculative theories presented at this meeting by amateur
     naturalists. In effect, the professionals delineated
     the boundaries of acceptable science.

241. Reingold, Nathan. "Definitions and Speculations: The
                  Professionalizations of Science in America in
                  the Nineteenth Century." *The Pursuit of Knowl-
                  edge in the Early American Republic: American
                  Scientific and Learned Societies from Colonial
                  Times to the Civil War* (item 209), pp. 33-69.

    Suggests a tripartite division of the American
scientific community to replace the traditional split
into amateurs and professionals: researchers, characterized by single-minded devotion to research, irrespective
of their vocational status; practitioners, who were paid
to engage in scientific or science-related activities
and whose research was distinctively less significant,
qualitatively and quantitatively, than researchers; and
cultivators, the lovers of science who had no vocational
ties to it.

PSEUDOSCIENCE

242. Butler, Jon. "Magic, Astrology, and the Early American
                  Religious Heritage, 1600-1760." *The American
                  Historical Review*, 84 (1979): 317-346.

    Argues that occult practices were a common form
of religious activity in the colonies prior to 1720,
having been brought from England. Views colonial almanacs
as an important source of occult information. Blames
the subsequent decline of interest partly upon opposition
from English and colonial elites, making the occult un-
fashionable and difficult to practice.

243. Davies, John D. *Phrenology, Fad and Science: A 19th-
                  Century American Crusade*. New Haven: Yale
                  University Press, 1955. xv + 203 pp.
                  Appendix, Bibliography, Index.

    Finds that the concept was imported from Europe
in the 1820s and 1830s by the intellectual community.
Discusses the transformation of the concept in the
United States into a highly popular applied science
used to improve the individual both physically and
morally. Traces the interaction of phrenology with
the theories and practices of education, the treatment
of the insane, penology, health reform, literature,
mesmerism, and medicine. Argues that the fiercest
opposition came from evangelical religious leaders
because phrenology was highly rationalistic and rejected
the concept of the depravity of man. Concludes that the

optimistic outlook of phrenology, with its belief in
the perfectability of man, struck a sympathetic cord
in Jacksonian America.

244. Leventhal, Herbert. *In the Shadow of the Enlightenment:
     Occultism and Renaissance Science in Eighteenth-
     Century America*. New York: New York University
     Press, 1976. 330 pp. Bibliography, Index.

     Argues that elements of Renaissance or Scholastic
world views, including belief in the occult (astrology,
witchcraft, and alchemy), Ptolemaic astronomy, and the
Aristotelian theories of the four qualities and four
elements, survived in America well into the eighteenth
century. Traces the subsequent decline of such beliefs.
Demonstrates a correlation between adherence to such
ideas and social class.

245. McVaugh, Michael, and Seymour H. Mauskopf. "J.B. Rhine's
     Extra-Sensory Perception and its Background
     in Psychical Research." *Isis*, 67 (1976):
     161-189.

     Describes *Extra-Sensory Perception* as a coherent
synthesis of a half century of psychical research rein-
forced by a new set of experiments which had achieved an
unprecedented success rate.

246. Moore, R. Laurence. "Spiritualism and Science: Reflections
     on the First Decade of the Spirit Rappings."
     American Quarterly, 24 (1972): 474-500.

     Argues that spiritualism gained rapid popularity
during the 1850s because it disassociated itself from the
occult, choosing instead to identify with empirical
science, especially the still-mysterious electrical
phenomena. Observes that the spiritualists were generally
opposed on religious, not scientific, grounds.

247. Stern, Madeleine B. *Heads and Headlines: The Phrenological
     Fowlers*. Norman: University of Oklahoma Press,
     1971. xx + 348 pp. Index.

     Discusses the activities of the Fowler family,
who successfully presented phrenology as a cure for the
ills of society through the reform of the individual.
Finds that the family remained active until the twentieth
century, when phrenology was displaced by psychoanalysis.

248. Stoehr, Taylor. *Hawthorne's Mad Scientists: Pseudo-
        science and Social Science in Nineteenth-
        Century Life and Letters.* Hamden, Conn.:
        Archon Books, 1978. 313 pp. Index.

        Contends that Hawthorne was "immersed" in the
pseudosciences of mesmerism, phrenology, and homeopathy,
and the activist, reformist social sciences of associa-
tionism, spiritualism, feminism, and prison reform.
Demonstrates their role in his writings. Argues that
he merged the two main strands of the mad scientist in
literature: the crackpot seeking utopian solutions and
the seeker after truths which should remain unknown
(e.g., Frankenstein).

249. Walsh, Anthony A. "The American Tour of Dr. Spurzheim."
        Journal of the History of Medicine, 27 (1972):
        187-205.

        Describes the brief 1832 visit of one of the
leading European phrenologists. Suggests that the major
contribution of the tour was the raising of public
awareness of phrenology.

250. Wilkinson, Ronald Sterne. "New England's Last Alchemists."
        Ambix, 10 (1962): 128-138.

        Shows how alchemical activity persisted in New
England through the first quarter of the nineteenth
century. Finds that most of the last generation of
alchemists were college graduates but were not familiar
with recent chemical discoveries.

SCIENCE AND RELIGION

251. Andrews, William D. "The Literature of the 1727 New
        England Earthquake." *Early American Literature*,
        7 (1973): 281-294.

        Finds that scientific explanations of the
earthquake were subordinated to theological interpretations.

252. Bodemer, Charles W. "Natural Religion and Generation
        Theory in Colonial America." *Cleo Medica*,
        11 (1976): 233-243.

        Identifies instances where the scientific
orientation of colonial thinkers regarding generation
and reproduction were influenced by the concepts of

natural religion. Cotton Mather and James Logan were
the two major examples.

253. Bozeman, Theodore Dwight. *Protestants in an Age of*
             Science: The Baconian Ideal and Antebellum
             American Religious Thought. Chapel Hill:
             The University of North Carolina Press, 1977.
             xv + 243 pp. Bibliography, Index.

        Focuses on the influence of science on the Old
School branch of the Presbyterian Church, which attempted
to reconcile Protestantism and science through the
Baconian philosophy. Finds that the Baconian methodology
enabled Presbyterians to embrace science in general, and
yet reject the more speculative aspects of science which
conflicted with their religious beliefs. Concludes that
later attacks on Darwin were not due to anti-intellectual-
ism but were the result of the antebellum decisions
regarding proper scientific methodologies.

254. ———. "Science, Nature and Society: A New Approach to
             James Henley Thornwell." *Journal of Presbyterian*
             History, 50 (1972): 306-325.

        Rejects the thesis that the leading theologian
in the ante-bellum South, a representative of the Old
School Presbyterian theology, was a product of conservative Biblicalism and pro-slavery ideology. Argues that
Thornwell's chief intellectual debt was to Presbyterian
rational orthodoxy, with its acceptance of natural
theology and the ultimate harmony of scientific and
religious views.

255. Fleming, Donald. *John William Draper and the Religion*
             of Science. Philadelphia: University of
             Pennsylvania Press, 1950. Reprint. New York:
             Octagon Press, 1972. ix + 205 pp. Bibliogra-
             phy, Index.

        Examines Draper's contributions to chemistry,
photography, evolutionary thought, and what has become
the traditional view of the conflict between science
and religion. Argues that Draper was one of the leading
American contributors to basic research during the
antebellum years.

256. Guralnick, Stanley M. "Geology and Religion Before
        Darwin: The Case of Edward Hitchcock,
        Theologian and Geologist (1793-1864)."
        Isis, 63 (1972): 529-543.

   Contends that Hitchcock's initial efforts to
reconcile Biblical and geological records were representative of the Anglo-American scientific communities of
that period. Discusses Hitchcock's gradual realization
that arguments supporting the equivalency of the truths
of Scripture and science were unnecessary for science
and irrelevant to religion.

257. Hovenkamp, Herbert. *Science and Religion in America,*
        1800-1863. Philadelphia: University of
        Pennsylvania Press, 1978. xii + 273 pp.
        Bibliography, Index.

   Examines the efforts of Protestant theologians
to establish a science of religion characterized by the
Baconian methodology, thus reconciling the scientific
and theological approaches to truth. Discusses their
eventual realization that laying the foundation of a
religion upon facts rather than faith would ultimately
be counterproductive. Explores, less successfully,
attempts by the scientific community to reconcile
knowledge with theology.

258. Lawrence, Philip J. "Edward Hitchcock: The Christian
        Geologist." *Proceedings of the American*
        Philosophical Society, 116 (1972): 21-34.

   Discusses Hitchcock's efforts to reconcile
science and religion through natural theology. For a
contrasting interpretation, see item 256.

259. Schoepflin, Gary Lee. "Denison Olmsted (1791-1859),
        Scientist, Teacher, Christian: A Biographical
        Study of the Connection of Science with
        Religion in Antebellum America." Ph.D.
        dissertation, Oregon State University, 1977.

   Offers a case study of a religious scientist--
a physical scientist on the Yale faculty with an explicit
concern for natural theology. Finds no direct links
between his choice of research topics and his religious
orientation. Concludes that religion provided a
motivation for doing science.

260. Sherwood, Morgan B.   "Genesis, Evolution, and Geology
                in America Before Darwin: The Dana-Lewis
                Controversy, 1856-1857." *Towards a History
                of Geology*. Edited by Cecil J. Schneer.
                Cambridge, Mass. and London: The M.I.T.
                Press, 1969, pp. 305-316.

     Examines Tayler Lewis's contention that it was
improper to use science to interpret the Bible and James
Dwight Dana's response that science was a unique and
powerful source of Biblical exegesis.

261. White, Edward A.   *Science and Religion in American
                Thought: The Impact of Naturalism*.   Stanford:
                Stanford University Press, 1952.   viii + 117 pp.

     Analyzes the positions taken by leading American
thinkers during the late nineteenth and early twentieth
centuries on the relationship between science and religion, which ranged from J.W. Draper's and Andrew White's
assaults against the clergy and the institutional power
of religion, through John Dewey's rejection of the
supernatural and reliance upon the natural world for
ultimate values, to the resurgence of the conflict in
the 1920s as the Fundamentalists raised the question
whether a science-based life could be moral.   Concludes
that the debate over this relationship has centered on
the nature of man, not Scripture.

SCIENCE, TECHNOLOGY, AND EDUCATION

262. Bell, Whitfield J., Jr.   "Thomas Cooper as Professor of
                Chemistry at Dickinson College, 1811-1815."
                *Journal of the History of Medicine and Allied
                Sciences*, 8 (1953): 70-87.

     Views Cooper's years at Dickinson as precipi-
tating a conflict between those who saw college as a
training ground for clergy and teachers of moral
principles and those who wanted higher education to
emphasize reason, science, and inquiry.

263. Ben-David, Joseph.   "The Universities of the Growth of
                Science in Germany and the United States."
                Minerva, 7 (1968-1969): 1-35.

     Argues that the American graduate school was
modeled upon inaccurate perceptions of the German university as a site for preparing students who possessed first
degrees for professional careers in research.

264. Broderick, Francis L. "Pulpit, Physics, and Politics:
    The Curriculum of the College of New Jersey,
    1746-1794." *William and Mary Quarterly*, 3d
    series, 6 (1949): 42-68.

    Discusses the raising of standards, the gradual
    secularization of purpose, and the increasing interest
    in science during Princeton's first six presidencies.
    Fails, however, to demonstrate the interrelationship
    of these three trends.

265. Chittenden, Russell. *History of the Sheffield Scientific
    School of Yale University, 1846-1922*. New
    Haven: Yale University Press, 1928. 2 Volumes.
    x + 610 pp. Appendix, Index.

    Presents a detailed history of the center for
    scientific, agricultural, and engineering studies at
    Yale. Credits Sheffield's ability to attract private
    donations as the key in its expansion from a school of
    applied chemistry. Concluded the study with the 1919
    reorganization, which changed the status of Sheffield
    from a separate entity to an integrated division of Yale.

266. Cohen, I. Bernard. *Some Early Tools of American Science:
    An Account of the Early Scientific Instruments
    and Mineralogical and Biological Collections
    in Harvard University*. Cambridge, Mass.:
    Harvard University Press, 1950. xxi + 201 pp.
    Appendices, Bibliography, Index.

    Provides detailed descriptions of the apparatus
    and collections. Contends that Harvard's holdings were
    the best in colonial North America and equal to those of
    many European universities. Emphasizes the difficulty
    in obtaining apparatus in the United States well into
    the nineteenth century. Notes but cannot satisfactorily
    explain the disappearance of botany from the curriculum
    during most of the eighteenth century.

267. Fleming, Donald. *Science and Technology in Providence,
    1760-1914: An Essay in the History of Brown
    University in the Metropolitan Community*.
    Providence: Brown University, 1952. 54 pp.
    Appendix.

    Provides an overview of the place of science
    and technology in Brown's curriculum. Argues that Way-
    land's curriculum reforms of the 1850s were an attempt
    to tie the college to Providence's industrial activities.

Finds little evidence of any interaction prior to
Wayland's efforts.

268. Guralnick, Stanley M.   "The American Scientist in Higher
          Education, 1820-1910." *The Sciences in the
          American Context: New Perspectives* (item 84),
          pp. 99-141.

          Divides the era into three periods: the expan-
sion of the curriculum by the addition of science courses
characterized the years 1820-1845; from 1845 through
1875 there were attempts at finding alternatives to the
required curriculum, including the establishment of
scientific schools; the period 1875-1910 was marked by
the grafting of the ideal of research to the other ideals
of education (intellectual and vocational training),
producing the modern institutional arrangement.

269. ————.   *Science and the Ante-Bellum American College*.
          Memoirs of the American Philosophical Society,
          volume 109. Philadelphia: American Philosophical
          Society, 1975. xiv + 227 pp. Appendices, Index.

          Finds that science permeated the curriculum of
these institutions. Provides detailed descriptions of
changes in textbooks, apparatus, and content in mathematics, astronomy, physics, and chemistry courses,
demonstrating the qualitative improvement and the
quantitative expansion. Includes biographical sketches
of the science faculty at the fifteen Northeastern
colleges used as a sample by this study.

270. ————   "Sources of Misconception on the Role of Science
          in the Nineteenth-Century American College."
          Isis, 65 (1974): 352-366.

          Examines and systematically rejects the tra-
ditional historiography which describes American colleges
as narrowly sectarian, classical, aristocratic, and
anti-science.

271. Hornberger, Theodore. *Scientific Thought in the American
          College, 1638-1800*. Austin: University of
          Texas Press, 1945. 108 pp. Index.

          Contends that science played a very small role
in colonial colleges prior to 1740, but subsequently it
was a major element in the curriculum, representing
20 percent to 40 percent of the coursework. Finds that
there was almost no field or laboratory work; teaching

was by lecture and recitation. Discusses the content for individual subjects.

272. McKeehan, Louis W. *Yale Science: The First Hundred Years, 1701-1801.* New York: Henry Schuman, 1947. x + 82 pp. Index.

Finds that science increased its share of the curriculum, but there was no research conducted; science was textbook learning, not experimentation.

273. Sinclair, Bruce. "The Promise of the Future: Technical Education." *Nineteenth-Century American Science: A Reappraisal* (item 32), pp. 249-272.

Finds that technical education was originally viewed as a means for working-class advancement and the easing of social distinctions. Observes that by the end of the century this democratic ideology had been replaced by the profit motive--technical education produced the technicians necessary for modern industry.

274. Stephens, Michael D., and Gordon W. Roderick. "American and English Attitudes to Scientific Education During the Ninteenth-Century." *Annals of Science*, 30 (1973): 435-456.

Credits America's economic leadership in the wake of the Second Industrial Revolution to its educational policies which resulted in a largely literate population familiar with some science, and an economic and political leadership fairly well versed in science and technology. Uses Massachusetts as an example.

275. Uberti, John Richard. "Men, Manners and Machines: The Young Man's Institute in Antebellum Philadelphia." Ph.D. dissertation, University of Pennsylvania, 1977.

Studies the Young Man's Institutes, organizations established by the financial community to encourage self-help, self-cultivation, and sobriety among mechanics. Focuses on the period 1850-1876. Argues that the objective of these organizations was the preparation of a new generation of mechanics for the challenges and requirements of the industrial workplace.

276. Weiner, Charles. "Science and Higher Education."
       Science and Society in the United States
       (item 99), pp. 163-189.

       Concludes that science has always been an
    integral part of the curriculum although the relative
    position has shifted in response to changes both in the
    content and organization of science and the structure and
    objectives of education.

SCIENCE, TECHNOLOGY, AND GOVERNMENT

277. Aldrich, Michele L. "American State Geological Surveys,
       1820-1845." *Two Hundred Years of Geology in
       America: Proceedings of the New Hampshire
       Bicentennial Conference on the History of
       Geology* (item 461), pp. 133-143.

       Demonstrates the central role of these organi-
    zations in the development of American geology.

278. ————. "New York Natural History Survey, 1836-1845."
       Ph.D. dissertation, University of Texas at
       Austin, 1974.

       Examines and rejects the thesis that there were
    significant differences in the treatment of the survey
    by the two major political parties. Concludes that the
    survey served as a model for other state surveys.

279. Allard, Dean Conrad, Jr. *Spencer Fullerton Baird and
       the U.S. Fish Commission*. New York: Arno
       Press, 1978. xii + 424 pp. Bibliography.

       Argues that Baird, the Fish Commissioner from
    1871 until his death in 1887, was largely responsible
    for the "golden age of government science" (the last
    quarter of the nineteenth century), and presents the
    Fish Commission as a typical government science agency.
    Shows the Fish Commission under Baird to have been
    characterized by the defense of basic research as
    necessary for the successful conduct of its mission,
    a sincere belief in the value of applied research, and
    strong political and public relations campaigns to
    ensure continued funding.

280. Allison, David Kite. "The Origin of Radar at the Naval
       Research Laboratory: A Case Study of Mission-
       Oriented Research and Development." Ph.D.
       dissertation, Princeton University, 1980.

Argues that an understanding of the history of radar requires an understanding of the research and development facility responsible. Provides an analysis of the Naval Research Laboratory prior to World War II, concentrating on the relationship of the Laboratory to the Bureau of Engineering, the struggle to maintain the Laboratory as a research center rather than simply an engineering facility, and the debate over whether the navy should conduct its own research, or contract it out to private industry.

281. Auerbach, Lewis E. "Scientists in the New Deal: A Pre-War Episode in the Relations Between Science and Government in the United States." *Minerva*, 3 (1965): 457-482.

Views the Science Advisory Board's brief existence (1933-1935) as an indication of the initial inability of the American scientific community to adjust to a situation in which the government sought its assistance and offered support. Contends that the Board's greatest contribution was the experience it provided the generation which would administer American science during World War II.

282. Baker, Gladys L., Wayne D. Rasmussen, Vivian Wiser, and Jane M. Porter. *Century of Service: The First 100 Years of the United States Department of Agriculture*. Washington, D.C.: U.S. Department of Agriculture, 1963. xvi + 560 pp. Appendix, Index.

Represents the official history of the department. Concentrates on the administrative aspects of the department and its responses to changing political, economic, and agricultural conditions.

283. Bartlett, Harley Harris. "The Reports of the Wilkes Expedition, and the Work of the Specialists in Science." *Proceedings of the American Philosophical Society*, 82 (1940): 601-705.

Summarizes and evaluates the scientific reports generated by the United States Exploring Expedition. Views the production of these reports as marking the entry of America into full partnership in world science.

284. Bartlett, Richard A. *Great Surveys of the American West*.
                Norman: University of Oklahoma Press, 1962.
                xxiii + 408 pp. Bibliography, Index.

         Describes the activities and accomplishments
of the Hayden, King, Powell, and Wheeler Surveys during
the 1860s and 1870s. Analyzes their policies towards
science, political lobbying, publicity, and surveying.
Concludes with their merger into the United States
Geological Survey.

285. Baxter, James Phinney, 3rd. *Scientists Against Time*.
                Boston: Little, Brown and Company, 1946.
                Reprint. Cambridge, Mass. and London: The
                M.I.T. Press, 1968. xxiv + 473 pp.
                Appendices, Index.

         Presents the official history of the Office of
Scientific Research and Development, the World War II
agency established to coordinate American defense research.
Provides an overview of American contributions to weapons
research and development, chemistry and chemical
warfare, military medicine, psychological testing for
the purposes of selection and training, and the atomic
bomb. For a companion history, focusing on the administrative side of the OSRD, see item 364.

286. Bruce, Robert V. *Lincoln and the Tools of War*.
                Indianapolis and New York: The Bobbs-Merrill,
                Company, 1956. xi + 368 pp. Bibliography,
                Index.

         Credits Lincoln with pressing for the adoption
of new military technology, such as breechloaders and
machine guns, over the opposition of conservative
military leaders. Contends that the military was more
concerned with ensuring sufficient quantities of quality
weapons and standardization than with innovation. Con-
cedes that in some areas, like artillery, the military
was probably correct in their conservatism. Argues
that the fundamental problem was the incompatibility of
research and production. The lack of an independent
research bureau within the government created a vacuum
which the ordnance bureaus, already preoccupied with
the problems surrounding the production of arms, could
only partially fill.

287. Buhl, Lance C.   "Mariners and Machines: Resistance to
                Technological Change in the American Navy,
                1865-1869." *The Journal of American History*,
                61 (1974): 703-727.

        Reevaluates the resistance of the post-Civil
War Navy to technological change and concludes that it
was based in part on good technical grounds: steam
propulsion was still relatively inefficient and expensive
compared to sail. In part, however, the resistance was
an effort by the line officers to protect their status
against the claims of engineering officers. The
adoption of the technology would have had social
consequences.

288. Burke, John G.   "Bursting Boilers and the Federal Power."
                Technology and Culture, 7 (1966): 1-23.

        Argues that the steam engine--a new technology
producing novel conditions--was pivotal in developing
public support for federal restrictions on the use of
private property where the safety of the public was
endangered. Traces the evolution of the regulatory
powers of the federal government between 1816 and 1852.

289. Christman, Albert B.   *Sailors, Scientists, and Rockets:*
                Origins of the Navy Rocket Program and of the
                Naval Ordnance Test Station, Inyokern.
                History of the Naval Weapons Center, China
                Lake, California. Volume I. Washington, D.C.:
                Government Printing Office, 1971. xxiii +
                303 pp. Appendices, Bibliography, Index.

        Provides the background for the Navy's program
of rocketry research and describes the first few months
of the Test Station's existence. Shows how World War I
taught the Navy the need for directed research, close
relationships between civilian and military experts,
interaction between scientists and engineers, and
sufficient lead time for developing a weapons system.
Describes the relationship between Naval ordnance and
the scientific community after the Armistice, however,
as one of mutual indifference, overcome only in the
wake of the general scientific mobilization for World
War II. Credits the adoption of the California Institute
of Technology's program on rockets by the Navy, which
marked the beginning of the Naval rocket program, to
their mutual bias for solid propellants.

290. Cochrane, Rexmond C. *Measures for Progress: A History of the National Bureau of Standards.* Washington, D.C.: Department of Commerce, 1966. Reprint. New York: Arno Press, 1976. xxv + 703 pp. Appendices, Bibliography, Index.

Views the founding of the bureau in 1901 as a response to both the growth of American industry and nationalistic attitudes that European nations should not be the only countries with such facilities. Discusses the expansion of the responsibilities of the bureau beyond the original conception of the establishment of standards and the determination of the properties of matter to the evaluation of the quality and performance of specific industrial and commercial materials destined for government use. Emphasizes the role of Samuel Wesley Stratton, the first director, in involving the bureau in industrial research problems, consumer fraud exposure, and the evaluation of manufactured goods. Contends that the concern with basic research rose and fell according to the attitudes of the director and the Executive Branch of the government.

291. Curti, Merle, and Kendall Birr. *Prelude to Point Four: American Technical Missions Overseas, 1838–1938.* Madison: The University of Wisconsin Press, 1954. xi + 284 pp. Bibliography, Index.

Studies government-sponsored efforts to export American technical knowledge. Finds no organized efforts prior to the Civil War, and little beyond agricultural missions during the rest of the nineteenth century. Notes a massive quantitative expansion during the twentieth century as American foreign policy became more expansive. Observes that most of the effort was aimed at Latin America and the Far East, where American interest was traditionally the highest. Identifies a number of variables which determined success, including American preconceptions, the threat to native vested interests, the language limitations and attitudes of American participants, and the nature of the technical knowledge.

292. Dupree, A. Hunter. "Central Scientific Organisation in the United States Government." *Minerva,* 1 (1963): 453–469.

Argues that every generation of American scientists has attempted to find a central focus to counterbalance the prevalent pluralism. Identifies two broad types of unifying foci: the explicit central scientific organization, which gives cohesion to all science, and the predominate agency, which has proven in practice to be more significant. The latter is a government bureau with a scientific mission of limited scope which temporarily expands its activities, eventually fulfilling the function of a central organization.

293. ———. "The *Great Instauration* of 1940: The Organization of Scientific Research for War." *The Twentieth-Century Sciences: Studies in the Biography of Ideas*. Edited by Gerald Holton. New York: W.W. Norton and Company, 1972, pp. 443-467.

Examines the transformation of the organization of American science through the agency of the Office of Scientific Research and Development. Argues that the objective of the OSRD was the application of science to military needs without disturbing the basic structure of American science.

294. ———. *Science in the Federal Government: A History of Policies and Activities to 1940*. Cambridge, Mass.: Harvard University Press, 1957. Reprint. New York: Arno Press, 1980. x + 460 pp. Chronology, Bibliography, Index.

Argues that the support provided by the federal government had been a very significant source of strength for American science from the very beginning. Provides the standard survey of that support. Finds that antebellum science-government relationships demonstrated five trends: practical problems reached out to theoretical considerations; *ad hoc* organizations became permanent; surveying and exploring were the key activities; great men were important; the quantity of scientific men increased. Observes that the post-Civil War period was characterized by the rise of civilian agencies and the decline of scientific activity in the military services. Concludes that the governmental science structure was firmly established by 1916 in the form of decentralized bureaus, confirmed by organic acts, and relatively free of political patronage pressures.

295. ——— "The Structure of the Government-University
Partnership After World War II." *Bulletin
of the History of Medicine*, 39 (1965):
245-251.

Contends that the active partnership between
universities and the federal government worked out after
World War II, enabling the universities to aid the nation
without losing their autonomy, represented "one of the
great social inventions in American history." Raises
issues and questions for further research.

296. England, J. Merton. "Dr. Bush Writes a Report: 'Science
--The Endless Frontier.'" *Science*, 191 (1976):
41-47.

Describes the process which led to the report
which laid the foundation for post-World War II science
policy. Finds that the critical element of the report--
the recommendation of a science foundation run by
Presidential appointees selected from National Academy
nominees--was a compromise between the scientists'
desire to be free of government oversight and the
necessity of some public control over public monies.

297. Ezrahi, Yaron. "The Political Resources of American
Science." *Science Studies*, 1 (1971): 117-133.

Argues that the dependence of American science
on outside financial support compels it to compete for
a share of public resources and political support.
Identifies four categories of lay perception of science:
relation of views of nature to prevailing social, political, and religious views; relation of scientifically
generated technology to prevailing social values;
accessibility of science to public understanding; peer
consensus among scientists.

298. Ferguson, Walter Keene. *Geology and Politics in Frontier
Texas, 1845-1909*. Austin and London:
University of Texas Press, 1969. xii + 233
pp. Appendices, Bibliography, Index.

Studies the relationship of science to the
political system in Texas from the founding of the first
geological survey until the establishment of the Bureau
of Economic Geology. Finds that political considerations
were important in staffing decisions; such considerations
also determined, in some cases, the direction of the
research. Views the Bureau of Economic Geology as a

successful effort to move geological research outside
the political arena. Admits that unique political and
economic problems, the lack of federal lands, the tremendous diversity in the state's climate, and the myth
of vast deposits of mineral wealth made Texas's history
atypical.

299. Frazier, Arthur H. *United States Standards of Weights
         and Measures: Their Creation and Creators*.
         Smithsonian Studies in History and Technology,
         No. 40. Washington, D.C.: Smithsonian Insti-
         tution Press, 1978. iii + 21 pp. Index.

Focuses on the essential activities of Joseph
Saxton, Ferdinand Hassler, and A.D. Bache.

300. Galtsoff, Paul S. *The Story of the Bureau of Commercial
         Fisheries Biological Laboratory, Woods Hole,
         Massachusetts*. United States Department of
         the Interior Circular 145. Washington, D.C.:
         Department of the Interior, 1962. iii + 121
         pp. Bibliography.

Surveys the history of the laboratory from its
founding in 1871 as the base of field operations for
the Commission of Fish and Fisheries to 1960. Identifies
its prime responsibility as basic research in support of
North Atlantic commercial fishing. Observes that the
laboratory has prospered with the exception of the years
just prior to and during World War II, when the Bureau
of Fisheries temporarily reversed its earlier supportive
attitude towards basic research.

301. Gilbert, James B. "Anthropometrics in the U.S. Bureau of
         Education: The Case of Arthur MacDonald's
         'Laboratory.'" *History of Education Quarterly*,
         17 (1977): 169-195.

Examines MacDonald's efforts to gather data
proving a direct link between physical appearance and
social undesirability (criminality, insanity, poverty)
while on the staff of the Bureau of Education. Concludes
that MacDonald's attempts to harmonize visual prejudices
and scientifically determined states of inferiority had
wide popular support.

302. Gilpin, Robert. *American Scientists and Nuclear Weapons Policy*. Princeton: Princeton University Press, 1962. ix + 352 pp. Index.

     Studies the scientist as a political animal, focusing on the "vocal," effective members of the scientific community in the determination of American policy towards nuclear weapons. Details the emergence, development, temporary victory, and subsequent failure of the "First Step" philosophy, which had contended that arms control could be a first step towards eventual disarmament. Argues that both the scientific community and the political leadership have failed to fully comprehend the relationships between the political and the technical realms when dealing with such an issue as the detection of nuclear tests.

303. Goetzmann, William H. *Army Exploration in the American West, 1803-1863*. New Haven: Yale University Press, 1959. xx + 509 pp. Bibliography, Index.

     Analyzes the role of the Army in Western exploration from Lewis and Clark until the Civil War, focusing on the contributions of the Corps of Topographical Engineers (established in 1838 and abolished in 1863). Describes the Corps as a small, elite organization which identified with the international scientific community. Emphasizes the scientific contributions of the Corps during surveys and exploring expeditions.

304. Goldberg, Steven. "The Constitutional Status of American Science." *University of Illinois Law Forum*, 1979, pp. 1-33.

     Argues that the Constitution guarantees the freedom of scientific activity through the guarantees of freedom of the press and speech, as well as the refusal to establish a religion. Contends further that the Constitution permits, and in some cases, requires governmental support for science through the establishment of certain governmental functions.

305. Harding, T. Swann. *Two Blades of Grass: A History of Scientific Development in the U.S. Department of Agriculture*. Norman: University of Oklahoma Press, 1947. Reprint. New York: Arno Press, 1980. xv + 352 pp. Appendix, Index.

Examines the practical returns from basic research conducted by the Department of Agriculture, and defends this research as highly cost effective. Divides the discussion according to themes rather than chronology; i.e., agricultural chemistry, economic entomology, plant science, tree husbandry, soil analysis, home economics, the dairy industry, and agricultural engineering.

306. Haskell, Daniel C. *The United States Exploring Expedition, 1838-1842, and its Publications, 1844-1874.* New York: New York Public Library, 1942. Reprint. New York: Greenwood Press, 1968. xii + 188 pp. Appendices, Index.

Gives locations for both printed and manuscript sources. Offers a brief evaluation of the scientific efforts. Remains a useful starting point.

307. Hechler, Ken. *Towards the Endless Frontier: History of the Committee on Science and Technology, 1959-79.* Washington, D.C.: Government Printing Office, 1980. xxxvi + 1073 pp. Appendix, Bibliography, Index.

Traces the history of the House Committee established to maintain American preeminence in science and technology. Discusses the space program, international scientific cooperation, the adoption of the metric system, the National Science Foundation, transportation, and energy policies from the perspective of the Congress. Highlights the relationship between the Executive and Legislative branches, jurisdictional struggles within Congress, and the defense of science against attacks by unsympathetic Congressmen.

308. Hendrickson, Walter B. "Nineteenth-Century State Geological Surveys: Early Government Support of Science." *Isis*, 52 (1961): 357-371.

Examines one of the earliest forms of American governmental support of scientific activities. Concludes that the usual justifications were the doctrine of mercantilism--the state should aid the economic efforts of its people--and the perception that surveys would further public education.

309. Hewlett, Richard G. and Oscar E. Anderson, Jr. *The
           New World, 1939/1946*. A History of the
           United States Atomic Energy Commission:
           Volume I. University Park: The Pennsylvania
           State University Press, 1962. xv + 766 pp.
           Appendix, Index.

     Chronicles the development of the atomic bomb
and the post-war efforts to devise an atomic policy
which culminated in the establishment of the Atomic
Energy Commission in 1946. Discusses events from the
perspective of the highest appropriate level of the
federal government. Demonstrates the complex interaction
of science, technology, politics, and the human element
(e.g., the morale of construction workers) in the process
of building the atomic bomb.

310. ———, and Francis Duncan. *Atomic Shield, 1947/1952*.
           A History of the Atomic Energy Commission:
           Volume II. University Park: The Pennsylvania
           State University Press, 1969. Reprint.
           Washington, D.C.: U.S. Atomic Energy Commission,
           1972. xviii + 718 pp. Appendices, Index.

     Traces the history of the Atomic Energy
Commission from its first meeting until the detonation
of the first thermonuclear bomb. Argues that during this
period the AEC discarded its hope that the primary utilization of atomic energy would be peaceful; in the name
of national security the AEC turned atomic energy into
a military shield against the Russians. Provides parallel
discussions of basic research in atomic physics, the
application of atomic energy to weaponry, the role of
atomic energy in policy making, and the problem of
administrating the AEC.

311. ———, and Francis Duncan. *Nuclear Navy, 1946-1962*.
           Chicago and London: The University of Chicago
           Press, 1974. xv + 477 pp. Appendices,
           Bibliography, Index.

     Documents the role of Hyman Rickover in the
adoption of nuclear propulsion by the United States Navy
and the development of the first full-scale, civilian
nuclear generating plant. Concludes that Rickover's
successful management technique was based upon high
engineering standards, attention to detail, reliance
upon individuals for knowledge rather than systems or
organizations, and careful and clear delineation of
responsibility.

312. Hill, Forest G. "Formative Relations of American
           Enterprise, Government and Science." *Political
           Science Quarterly*, 75 (1960): 400-419.

     Assesses the role of the military establishment
in the antebellum transfer of technical knowledge from
Europe, the provision of technical training, and scientific
exploration. Contends that the military participated in
these activities both because it had responsibilities
which necessitated scientific knowledge and because
such activities helped diffuse public hostility to the
standing army.

313. ———. *Roads, Rails & Waterways: The Army Engineers
           and Early Transportation*. Norman: University
           of Oklahoma Press, 1957. xi + 248 pp.
           Bibliography, Index.

     Analyzes the impact of the technical expertise
contributed by Army engineers during the westward expansion of the United States between the War of 1812 and
the Civil War. Contends that the General Survey Act of
1824 was an effort to institute a national plan of
internal improvements. Argues that the planning en-
visioned by the 1824 Act never came to fruition, although
the Act did enable the military to supply technical
support for a wide range of civilian projects. Claims
that the termination of the 1824 Act in 1838 and the
resulting sharp decline in military participation in
civilian internal improvements was due to the rising
number of civilian engineers and the switch in emphasis
from roads and canals to railroads; the latter were
identified with private profits and local interests, and
were considered less favorable candidates for federal
support.

314. Hughes, Patrick. *A Century of Weather Service: A History
           of the Birth and Growth of the National Weather
           Service, 1870-1970*. New York, London, and
           Paris: Gordon and Breach, 1970. xii + 212 pp.
           Chronology, Index.

     Credits technology with playing an essential
role in the development of weather forecasting by making
it possible to gather observations rapidly from a wide
area and then disseminate the information to the public.
Focuses on the role of war in encouraging American efforts
in meteorology, especially the two World Wars, and the
contributions of meteorology to modern warfare.

315. Hughes, Thomas Parke. "Technology and Public Policy:
               The Failure of Giant Power." *Proceedings of
               the IEEE*, 64 (1976): 1361-1371.

               Argues that this regional electrification plan
     for Pennsylvania, proposed in 1923, was not rejected for
     technical reasons, but because it represented a radical
     approach and would bring about social change.

316. Jones, Kenneth Macdonald. "The Endless Frontier."
               Prologue, 8 (1976): 33-46.

               Analyzes the public's attitude towards science
     and scientists in the wake of the appearance of Vannevar
     Bush's *Science the Endless Frontier*. Finds a consensus
     that science had greatly aided the war effort, would
     continue to contribute to the well-being of the nation,
     and would require a government program of some form for
     the encouragement of research.

317. Kazar, John Dryden, Jr. "The United States Navy and
               Scientific Exploration, 1837-1860." Ph.D.
               dissertation, University of Massachusetts, 1973.

               Views these expeditions in terms of contrasting
     values: nationalism and national prestige versus the
     internationalism of science; democratic science, charac-
     terized by the wide, popular diffusion of knowledge
     versus the professional ideal of scholarly publication.
     Concludes that scientists found the support of the Navy
     "indispensable," despite conflicts and disagreements.

318. Kevles, Daniel J. "Federal Legislation for Engineering
               Experiment Stations: The Episode of World War
               I." *Technology and Culture*, 12 (1971): 82-189.

               Argues that the attempt during World War I to
     establish engineering experiment stations in the states
     was an important benchmark in the evolution of the rela-
     tionship between science and the federal government
     because it raised the issue of how federal funds should
     be distributed--on the basis of geography or the quality
     of the projects seeking support.

319. ———. "Flash and Sound in the AEF: The History of a
               Technical Service." *Military Affairs*, 33
               (1969): 374-384.

               Examines the role of science in spotting
     enemy artillery.

320. ———. "George Ellery Hale, the First World War, and
the Advancement of Science in America." *Isis*,
59 (1968): 427-437.

Describes Hale's attempts to encourage cooperative research and provide the prestige of government
patronage to American science which culminated in the
establishment of the National Research Council. Argues
that the Council's efforts after the war were restricted
by the limited financial support provided by the federal
government.

321. ———. "Hale and the Role of a Central Scientific
Institution in the United States." *The Legacy
of George Ellery Hale*. Edited by Joan N.
Warnow and Charles Weiner. Cambridge, Mass.
and London: The M.I.T. Press, 1972, pp.
273-282.

Summarizes Hale's efforts to promote cooperative,
interdisciplinary research through the National Academy
of Sciences and the National Research Council. Concludes
that Hale was seeking a compromise which would satisfy
the need for a central scientific institution, while
respecting the decentralized nature of American society.

322. ———. "The National Science Foundation and the Debate
over Postwar Research Policy, 1942-1945: A
Political Interpretation of *Science--The Endless
Frontier*." *Isis*, 68 (1977): 5-26.

Argues that Bush's report was a conservative
response to Sen. Harley M. Kilgore's vision of applying
America's scientific resources to the solution of social
and economic problems. Bush, in contrast, wanted to use
the social and economic resources of the nation to advance the cause of science. Concludes that the National
Science Foundation represented a victory for Bush.

323. ———. "Scientists, the Military, and the Control of
Postwar Defense Research: The Case of the
Research Board for National Security, 1944-46."
Technology and Culture, 16 (1975): 20-47.

Argues that this attempt at ensuring continued
civilian participation in postwar defense research programs floundered over the issue of democratic versus
elitist controls. The debate centered on the question
of whether the Board should be an independent federal
agency, subject to Congressional and Presidential

controls, or act as a distributor and coordinator of
civilian research funded by the military, while only
responsible to the National Academy of Sciences.

324. Lasby, Clarence G. "Science and the Military." *Science
     and Society in the United States* (item 99),
     pp. 251-282.

     Concludes that for most of this country's history
the relationship between science and the military, while
mutually fruitful, was based on expediency rather than
any natural affinity. The antebellum period, when the
two groups cooperated in exploration and the officer
ranks were filled with technically oriented men, was
followed by a half-century of mutual disassociation, as
officers evolved into professional soldiers. The twentieth
century has seen the rise of cooperative efforts at
weapons research.

325. Lenzen, V.F. *Benjamin Peirce and the U.S. Coast Survey.*
     San Francisco: San Francisco Press, 1968.
     vii + 54 pp. Appendices, Index.

     Provides a brief biographical sketch and analysis
of Peirce's scientific contributions. Fails to examine
the problems involved in administering a government
science bureau.

326. Lomask, Milton. *A Minor Miracle: An Informal History of
     the National Science Foundation.* Washington,
     D.C.: National Science Foundation, 1976.
     x + 285 pp. Index.

     Examines the quintessential role of the directors
in the establishment of the policies of the NSF. Focuses
on the rejection of the role of science czar by Alan
Waterman, the first director, despite pressure upon him
by other members of the government to assume that role.
Claims that Waterman was correct in limiting the policy-
making role of NSF and concentrating the agency's energies
upon the promotion of basic research and science education,
an orientation Waterman's successors continued. Identifies Project Mohole as the source of the first major
rift between the science community and NSF, contending
that the underlying problem was the agency's inability
to negotiate proper contracts and manage large-scale
projects. Summarizes the NSF's contributions to the
training of scientists. Discusses the debate over the
NSF's involvement in applied research.

327. Lundeberg, Philip K. *Samuel Colt's Submarine Battery:*
        The Secret and the Enigma. Smithsonian Studies
        in History and Technology, No. 29. Washington,
        D.C.: Smithsonian Institution Press, 1974.
        v + 90 pp. Appendices, Index.

    Discusses Colt's ultimately unsuccessful efforts
to develop a system of harbor defenses utilizing underwater mines detonated by electricity. Contends that
Colt's greatest contribution was his single observer
method of target acquisition. Argues that Colt's secretive
manner antagonized the military and eventually led to
the loss of government support.

328. McCraw, Thomas K. "Triumph and Irony--The TVA." *Pro-*
        ceedings of the IEEE, 64 (1976): 1372-1380.

    Contends that the core of the Tennessee Valley
Authority was the concept of multipurpose river development. Divides the history of the TVA into three phases:
1933-1941 was marked by the construction of multipurpose
dams and the quest for power markets; from 1941 through
1961 there was an increase in power generation to meet
increasing demand; since 1961 the sense of mission has
faded, while ecological and cost questions have arisen.

329. Manning, Thomas G. *Government in Science: The U.S.*
        Geological Survey, 1867-1894. Lexington:
        University of Kentucky Press, 1967. xiv +
        257 pp. Bibliography, Index.

    Examines the history of the Survey at a time
when it was the leading scientific bureau in the government, in an attempt to illustrate fundamental issues
regarding science in government. Finds a constant
tension between the military and the civilian scientific
community over control of governmental surveying. Discusses Congressional pressure upon the Survey for
practical knowledge. Blames John Wesley Powell's
problems as director upon his doctrinaire, anti-
democratic attitude which was incompatible with the
political climate within which the Survey operated.

330. ————. "Peirce, the Coast Survey, and the Politics of
        Cleveland Democracy." *Transactions of the*
        Charles S. Peirce Society, 1975 (11): 187-194.

    Illustrates the subversion of science by
politics. Documents the fiscal reforms and employment
of political appointees without scientific qualifications

which discouraged or limited research activities,
including C.S. Peirce's gravity studies, within the
Coast Survey.

331. Merrill, George P. *Contributions to a History of
              American State Geological and Natural History
              Surveys.* United States National Museum,
              Bulletin 109. Washington, D.C.: U.S. National
              Museum, 1920. Reprint. New York: Arno Press,
              1978. xviii + 549 pp. Appendix, Index.

         Discusses the surveys state by state. Approaches
    the surveys as administrative units, focusing upon
    legislative actions, appropriations, and expenditures.
    Remains the standard work.

332. Nash, Gerald D. "The Conflict Between Pure and Applied
              Science in Nineteenth-Century Public Policy:
              The California State Geological Survey, 1860-
              1874." *Isis,* 54 (1963): 217-228.

         Argues that disagreements over the objectives
    of the survey--basic research or practical results--led
    to its demise. Contends that divergent conceptions of
    the proper activities of government research agencies
    was a characteristic problem during the nineteenth
    century.

333. Nelson, Clifford M., and Ellis L. Yochelson. "Organizing
              Federal Paleontology in the United States,
              1858-1907." *Journal of the Society for the
              Bibliography of Natural History,* 9 (1980):
              607-618.

         Traces the continuing presence of paleontology
    in the federal government to the coming of Fielding B.
    Meek to the Smithsonian. Credits Charles D. Walcott
    with establishing the modern administrative framework
    for paleontology at both the Geological Survey and the
    Smithsonian.

334. Pearson, Lee M. "The 'Princeton' and the 'Peacemaker':
              A Study in Nineteenth-Century Naval Research
              and Development Procedures." *Technology and
              Culture,* 7 (1966): 163-183.

         Concludes that political authority was utilized
    by the developer of new technology to nullify technical
    criticism.

335. Pickard, Madge E. "Government and Science in the United States: Historical Backgrounds." *Journal of the History of Medicine and Allied Sciences,* 1 (1946): 254-289, 446-481.

    Views the Smithsonian as the first comprehensive national science agency. Examines the history of its unsuccessful predecessors, the Columbian Institute and the National Institute.

336. Plotkin, Howard. "Astronomers Versus the Navy: The Revolt of American Astronomers over the Management of the United States Naval Observatory, 1877-1902." *Proceedings of the American Philosophical Society,* 122 (1978): 385-399.

    Discusses the attempts by segments of the astronomical community to free the Naval Observatory from military control. Argues that it was part of a more general debate on the issue of military versus civilian control of governmental science bureaus.

337. Ponko, Vincent, Jr. *Ships, Seas, and Scientists: U.S. Naval Exploration and Discovery in the Nineteenth Century.* Annapolis: Naval Institute Press, 1974. xii + 283 pp. Bibliography, Appendix, Index.

    Surveys, without attempting a synthetic analysis, the Naval exploring expeditions conducted between 1838 and 1861. Blames the decline in number of such expeditions after the Civil War on the shift in American exploring emphasis from the sea to the West, the rise of civilian scientific exploring expeditions, and the increasing costs of expeditions utilizing the steam-powered boats of the post-war Navy.

338. Post, Robert C. "'Liberalizers' Versus 'Scientific Men' in the Antebellum Patent Office." *Technology and Culture,* 17 (1976): 24-54.

    Demonstrates that the ratio of patents to applications varied considerably according to the attitude of the patent examiners. Finds that it was especially low from 1842 through 1853 when the examiners were a corps of professional scientists with a very narrow definition of novelty. After 1853 the ratio climbed as a consequence of political pressure to replace the examiners with men with more liberal criteria for novelty.

339. ———. "The Page Locomotive: Federal Sponsorship of
              Invention in Mid-19th-Century America."
              Technology and Culture, 13 (1972): 140-169.

         Assesses Page's invention as neither a spin-off
    from a scientific discovery nor a response to a social
    need, but rather the premature efforts of a man who
    overestimated his own knowledge. Contends Page received
    federal funds because he was a skilled lobbyist who was
    able to enlist the support of a powerful Senator--
    Thomas H. Benton.

340. Pursell, Carroll W., Jr. "The Administration of Science
              in the Department of Agriculture, 1933-1940."
              Agricultural History, 42 (1968): 231-240.

         Contends that scientific research had a stronger
    position in the Department of Agriculture in 1940 than
    before the New Deal because of Secretary Wallace's
    support of science and willingness to exploit opportuni-
    ties offered by the New Deal.

341. ———. "The Farm Chemurgic Council and the United
              States Department of Agriculture, 1935-1939."
              Isis, 60 (1969): 307-317.

         Examines an attempt by the Chemical Foundation
    to solve the overproduction crisis facing American
    agriculture by searching for new industrial uses for
    farm products. Argues that the program had two major
    weaknesses: it did not provide any immediate relief to
    farmers and the number of new uses actually discovered
    was quite low.

342. ———. "Government and Technology in the Great Depres-
              sion." *Technology and Culture*, 20 (1979):
              162-174.

         Finds that the government's policy was to
    maximize the positive effects of technology, study the
    negative ones, and help the victims. Concludes it
    avoided fundamental questions regarding the role of
    technology in society, such as the dehumanization of
    man by the machine.

343. ———. "A Preface to Government Support of Research
              and Development: Research Legislation and the
              National Bureau of Standards, 1935-41."
              Technology and Culture, 9 (1968): 145-164.

Argues that efforts to gain Congressional approval for the Bureau to provide basic research for American industry failed in part because of worries over the cost and consequences of such a program, and in part because other organizations and governmental agencies wanted a share of the funds.

344. ———. "Science Agencies in World War II: The OSRD and Its Challengers." *The Sciences in the American Context: New Perspectives* (item 84), pp. 359-378.

Discusses the political and administrative infighting that accompanied the efforts by the Office of Scientific Research and Development to establish and maintain its position as the chief American science and technology agency during World War II. Contends that the OSRD represented a commitment to the conservative, prewar attitude towards the distribution and use of resources for science.

345. ———. "Science and Government Agencies." *Science and Society in the United States* (item 99), pp. 223-249.

Views the bureau as the most fruitful organizational form for governmental science. Characterizes it as having a specific mission mandated by Congress, consisting of a group of scientists working as a team, and able to rely on a definite outside constituency for political support.

346. Rabbitt, Mary C. *Minerals, Lands, and Geology for the Common Defense and General Welfare*. Volume I: *Before 1879*. Washington, D.C.: Government Printing Office, 1979. x + 331 pp. Bibliography, Indices.

Provides a chronological account of the history of the geological sciences in the United States prior to the establishment of the Geological Survey. Concentrates on the themes of public land use, federal science and mapping policies, and the development of mineral resources. Focuses on the connection of geology with the economic development of the United States.

347. ———. *Minerals, Lands, and Geology for the Common
          Defense and General Welfare*. Volume II:
          1879-1904. Washington, D.C.: Government
          Printing Office, 1980. vii + 407 pp.
          Bibliography, Indices.

     Provides a chronological account of the activities
of the United States Geological Survey during the direc-
torships of King, Powell, and Walcott. Focuses on the
same themes as item 346. Argues that the turn of the
twentieth century marked the point when America became
more concerned with preserving natural resources than
developing them. Contends that the turn of the century
was also a period of an acceleration of the bifurcation
of basic and applied research; the Geological Survey was
rapidly becoming a department of applied science. Concludes that one of the major achievements of the
Geological Survey during these twenty-five years was
its role in the development of economic geology as a
science.

348. Redmond, Kent C. "World War II, a Watershed in the Role
          of the National Government in the Advancement
          of Science and Technology." *The Humanities
          in the Age of Science*. Edited by Charles
          Angoff. Rutherford, Madison, and Teaneck:
          Fairleigh Dickinson University Press, 1968,
          pp. 166-180.

     Finds that the traditional governmental policy
consisted of ad hoc responses to acute problems, and the
establishment of bureaus or departments in response to
chronic problems or the needs of dominant economic or
political groups; there were no attempts to coordinate
the activities of the bureaus and no programs for
facilitating exchanges of information. Claims that
World War II brought home the intimate relationship
between science, technology, and the national welfare,
resulting in the establishment of agencies for the
coordination, direction, and financing of research and
development.

349. Rees, Mina S. "Mathematics and the Government: The
          Post-War Years as Augury of the Future." *The
          Bicentennial Tribute to American Mathematics,
          1776-1976*. Edited by Dalton Tarwater. n.p.:
          The Mathematical Association of America,
          1977, pp. 101-116.

Reviews the changes in university research
resulting from the large influx of government funding
via the Office of Naval Research. Contends that this
support was especially important for the development of
computer science and statistics.

350. Reingold, Nathan. "Science in the Civil War: The
        Permanent Commission of the Navy Department."
        Isis, 49 (1958): 307-318.

Argues that the Civil War was not a "scientific"
war--i.e., not a war characterized by the application of
science to weapon development and other military needs.
Finds that the lack of institutions to conduct research
and development led the armed forces to rely upon amateur
inventors. Describes the Permanent Commission's activities
in screening the suggestions of such inventors.

351. Rezneck, Samuel. "The Emergence of a Scientific Community
        in New York State a Century Ago." *New York
        History*, 43 (1962): 211-238.

Views the American political system as offering
both opportunity and circumscription to science. Characterizes American science as a mixture of the practical
and the theoretical as it attempts to satisfy both the
scientific community and the public. Presents the New
York State Natural History Survey as a typical antebellum
American scientific institution.

352. Sapolsky, Harvey M. "Academic Science and the Military:
        The Years Since the Second World War." *The
        Sciences in the American Context: New
        Perspectives* (item 84), pp. 379-399.

Examines the background and development of the
most important federal agency for the support of academic
science between World War II and the Korean War--the
Office of Naval Research. Asserts that it was very
influential in determining the subsequent university-
government relationships regarding campus-based research.

353. Schiff, Ashley L. *Fire and Water: Scientific Heresy in
        the Forest Service*. Cambridge, Mass.: Harvard
        University Press, 1962. xi + 224 pp.
        Appendix, Index.

Offers a case study of the relationship of a
government research unit to the administrative units
responsible for the application of the technical knowledge

generated by the research unit; the Forest Service was
an example of the integration of the researchers into
the administrative unit. Examines the conflict between
researchers and administrators on two issues: the impact
of fire on forests and forests on stream flow. Finds
that in both cases the Forest Service promoted its own
position despite mounting scientific evidence for the
opposite approach. Concludes that the administrative
structure of the Forest Service led to parochialism,
rigidity, and conformity.

354. Sherwin, Martin J. *A World Destroyed: The Atom Bomb and
         the Grand Alliance*. New York: Alfred A. Knopf,
         1975. xvi + 315 + xi pp. Appendices,
         Bibliography, Index.

Argues that Roosevelt's decision to restrict
knowledge of the atomic bomb was part of his long-range
plan for the Anglo-American partnership to act as global
policemen in the post-war world. Contends that this
decision was made without consulting his military and
scientific advisors. Views this as part of a distinctly
anti-Russian foreign policy.

355. Sherwood, Morgan B. *Exploration of Alaska, 1865-1900*.
         New Haven and London: Yale University Press,
         1965. xv + 207 pp. Index.

Tries to place the exploration of Alaska in
the contexts of the history of science, the history of
exploration, and the history of Western expansion. Shows
parallels with the pattern of exploration of the lower
forty-eight states one to two generations earlier.
Documents the scientific activities of the Army, Coast
Survey, Geological Survey, and the Smithsonian. Contends
that the lack of population and the perceived limited
possibilities for economic exploitation were the
reasons for the relatively slow and limited pace of
the exploration of the region.

356. ———. "Federal Policy for Basic Research: Presidential
         Staff and the National Science Foundation,
         1950-1956." *The Journal of American History*,
         55 (1968): 599-615.

Discusses the reluctance of the National
Science Foundation to embrace the opportunity to become
the policy-making, coordinating, and evaluating body for
American science. Credits this reluctance to the perception that the scientific community would oppose such
efforts by NSF.

357. ———. "Specious Speciation in the Political History
of the Alaskan Brown Bear." *The Western
Historical Quarterly*, 10 (1979): 49-60.

Illustrates the dangers when political policy
is based on inconclusive or debatable scientific knowl-
edge--in this case, the definition of a species.

358. Sinclair, Bruce. *Early Research at the Franklin Institute:
The Investigation into the Causes of Steam Boat
Explosions, 1830-1837*. Philadelphia: The
Franklin Institution, 1966. iii + 28 pp. +
Facsimile.

Describes the first outside research sponsored
by the federal government. Includes a facsimile of the
1837 General Report.

359. Smith, Alice Kimball. *A Peril and a Hope: The Scientists'
Movement in America, 1945-47*. Chicago and
London: The University of Chicago Press, 1965.
xiv + 591 pp. Appendices, Index.

Details the lobbying, educational, and organizational efforts of the American scientific community,
led by the atomic physicists, to ensure the control of
atomic energy and the free exchange of scientific information. Demonstrates the preoccupation with this issue
at the expense of other aspects of the post-war relationship
between science and the federal government. Relies
heavily on the recollections of the participants.

360. Smith, Geoffrey Sutton. "The Navy Before Darwinism:
Science, Exploration, and Diplomacy in Antebellum
America." *American Quarterly*, 28 (1976): 41-55.

Advocates treating the antebellum Naval exploration and support of scientific activity as part of
America's diplomatic efforts to secure a commercial
empire. Praises Matthew Fontaine Maury's efforts in
developing the Naval Observatory into a training ground
for naval scientists and a clearinghouse for information.

361. Smith, Maryanna S. *A List of References for the History
of the United States Department of Agriculture*.
Davis: Agricultural History Center, University
of California, 1974. iv + 88 pp. Index.

Provides some descriptive annotations and an
author index. Includes published material, dissertations,
and processed materials. Claims completeness.

362. Stanton, William. *The Great United States Exploring Expedition of 1838-1842*. Berkeley, Los Angeles, and London: University of California Press, 1975. x + 433 pp. Bibliography, Index.

Offers an analysis of the political, economic, and intellectual forces leading to the establishment of the Wilkes Expedition. Proves to be much stronger in its evaluation of the scientific contributions of the Expedition than item 365. Credits the Expedition with obtaining international respect for the American scientific community and legitimizing intellectual activity in the egalitarian American society.

363. Steiner, Arthur. "Scientists, Statesmen, and Politicians: The Competing Influences on American Atomic Energy Policy, 1945-46." *Minerva*, 12 (1974): 469-509.

Rejects the thesis that the scientific community was omniscient in its advice on these policy issues. Points out specific errors regarding international agreements, the place of the atomic bomb in weapons systems, and the future dominant strategic issues.

364. Stewart, Irwin. *Organizing Scientific Research for War: The Administrative History of the Office of Scientific Research and Development*. Boston: Little, Brown and Company, 1948. Reprint. New York: Arno Press, 1980. xiv + 358 pp. Appendix, Index.

Organizes the history under the rubrics of organization, liaison, supporting operations, and demobilization. Argues that success depended upon the efforts of the operating units, whose histories are told in separate volumes. Provides the perspective, generally, of the highest and most central authority. This official history was written by a former Deputy Director.

365. Tyler, David B. *The Wilkes Expedition: The First United States Exploring Expedition (1838-1842)*. Memoirs of the American Philosophical Society, 73. Philadelphia: American Philosophical Society, 1968. xvi + 435 pp. Appendix, Bibliography, Index.

Concentrates on the exploring and surveying activities of the expedition. Evaluates the performance of Charles Wilkes as commander, focusing on his relation-

ship with the officer corps and civilian scientists.
Item 362 is much superior in evaluating the scientific
activities and accomplishments of the expedition.

366. Useem, Michael. "Government Patronage of Science and
              Art in America." *American Behavioral Scientist*,
              19 (1975-76): 785-804.

    Examines four models of governmental support of
culture and tentatively concludes that the federal government funds research in the physical sciences because of
potential economic returns, in essence subsidizing big
business, and supports the social sciences because
government needs and utilizes their discoveries.

367. York, Herbert F. *The Advisors: Oppenheimer, Teller, and
              the Superbomb*. San Francisco: W.H. Freeman and
              Company, 1976. x + 175 pp. Appendix, Index.

    Explores the reasoning leading to the recommenda-
tion of the Oppenheimer-led General Advisory Committee
of the Atomic Energy Commission in 1949 that the United
States should forego development of the Superbomb in
favor of improved, more conventional atomic weapons.
Concludes that their evaluation of the state of nuclear
weapons technology was correct; adopting their recommendation would have left open the possibility of disarmament
without threatening the security of the United States.

SCIENCE, TECHNOLOGY, AND WOMEN

368. Cowan, Ruth Schwartz. "Women and Technology in American
              Life." *Technology at the Turning Point*. Edited
              by William B. Pickett. San Francisco: San
              Francisco Press, 1977, pp. 23-33.

    Claims that women have a different relationship
with technology than men. Offers such examples as the
uniquely female-oriented technologies dealing with child-
bearing and rearing, technologies linked to the sex-typing
of certain jobs, including household technology, and the
role of women as anti-technocrats.

369. Kohlstedt, Sally Gregory. "In from the Periphery:
              American Women in Science, 1830-1880." *Signs*,
              4 (1978-79): 81-96.

    Divides these women into three generations:
an initial generation characterized by individual and
generally unrecognized effort; a mid-century group of

educators and popularizers; a late-century generation
which challenged the system and entered the scientific
community. Argues that although the second generation
remained on the periphery, their very existence refuted
claims that women could not understand science and provided a foundation for the activities of the following
generation.

370. ———. "Maria Mitchell: The Advancement of Women in
         Science." *The New England Quarterly*, 51
         (1978): 39-63.

Examines Mitchell's efforts to increase the
visibility and self-confidence of female scientists as
well as encourage their aspirations.

371. Rossiter, Margaret W. "Women Scientists in America Before
         1920." *American Scientist*, 62 (1974): 312-323.

Concludes that peer recognition and career
possibilities were limited compared to their contemporary
male colleagues.

372. ———. "'Women's Work' in Science, 1880-1910." *Isis*,
         71 (1980) 381-398.

Argues that occupational sex segregation in
American science during this formative period took two
forms: a territorial kind in which the entire discipline
or field was sex-typed, such as home economics, and a
hierarchical kind in which women were relegated to low-
level positions, such as occurred in astronomy. Notes
that female employment in science was limited to fields
or positions where there was a shortage of men.

373. Trecker, Janice Law. "Sex, Science and Education."
         American Quarterly, 26 (1974): 352-366.

Contends that sexism permeated many of the
assumptions of nineteenth-century American scientists,
resulting in biased social theories. Accuses the scientifi
community of taking very conservative positions on the
issue of women's education to prevent professional and
economic competition, and to neutralize one of the most
effective weapons in the feminist arsenal--education.

374. Warner, Deborah Jean. "Science Education for Women in
            Antebellum America." *Isis*, 69 (1978): 58-67.

        Demonstrates that science became increasingly
    prominent in the curricula of women's schools, paralleling,
    although on a smaller scale, developments in men's
    colleges. Finds that by 1860 there was present in
    American society a supportive context for the participa-
    tion of women in science.

375. ————. "Women Astronomers." *Natural History*, 58,
            No. 5 (May 1979): 12-26.

        Traces the efforts of women to become an
    integral part of the American astronomical community.
    Finds that after their achieving apparent success by the
    1920s, disenchantment set in as they came to the realiza-
    tion that most positions open to women remained low-status
    ones and that a choice had to be made between career
    and marriage.

376. Wilson, Joan Hoff. "Dancing Dogs of the Colonial Period:
            Women Scientists." *Early American Literature*,
            7 (1973): 225-235.

        Contends that the previous historiography has
    either completely ignored female contributors to science
    or acknowledged only the rare exception whose accomplish-
    ments have met the standards imposed upon male scientists.
    Argues that the historian must borrow the "structural-
    functional" approach from the sociologist if female
    scientists are to be correctly evaluated.

377. Yost, Edna. *American Women of Science*. Philadelphia
            and New York: Frederick A. Stokes Company,
            1943. xviii + 232 pp.

        Presents biographical sketches of twelve
    female scientists from Ellen H. Richards to Margaret
    Mead, arranged chronologically. The goal was to present
    women with role models rather than to suppy historical
    insights.

CHAPTER III: THE PHYSICAL SCIENCES

ASTRONOMY

378. Bedini, Silvio A. *The Life of Benjamin Banneker*. New
York: Charles Scribner's Sons, 1972. xviii +
434 pp. Bibliography, Index.

Examines the life and work of the first Black
American to gain a reputation in the scientific sphere--
a landed Maryland freeman who was an almanac maker and
surveyor. Describes and evaluates Banneker's astronomical
computations. Gives a full explanation of the production
of almanacs.

379. Bell, Whitfield J., Jr. "Astronomical Observatories of
the American Philosophical Society, 1769-1843."
*Proceedings of the American Philosophical
Society*, 108 (1964): 7-14.

Describes the unsuccessful attempts by the
American Philosophical Society to establish an astronomical
observatory. Blames the failures on insufficient funds
and the ability of APS members to find alternatives.

380. Berendzen, Richard. "Origins of the American Astronomical
Society." *Physics Today*, 27 (December 1974):
32-39.

Analyzes the difficulties created during the
formation of this organization by the dichotomy between
the astrophysicists and traditional astronomers.
Delineates the contributions of George Ellery Hale
and Simon Newcomb.

381. Brush, Stephen G. "A Geologist Among Astronomers: The
        Rise and Fall of the Chamberlin-Moulton
        Cosmogony." *Journal for the History of
        Astronomy*, 9 (1978): 1-41, 77-104.

    Discusses the development and fate of the
theory proposed jointly by the geologist T.C. Chamberlin
and the astronomer F.R. Moulton that the planets accreted
from small solid particles formed during the encounter
of the Sun with another star. Finds that the history
of this theory provides two counterexamples to the model
of science developed by Imre Lakatos.

382. ————. "The Rise of Astronomy in America." *American
        Studies*, 20, No. 2 (1979): 41-67.

    Rejects Dupree's argument (see item 37) that
there is such an entity as American science, contending
that there is no evidence of a distinctly American
method of doing science. Identifies leading events
and individuals in the history of astronomy and tabulates
them by nationality. Finds no evidence of significant
astronomical activity in this country prior to 1850 but
the domination of the discipline by Americans during
the twentieth century.

383. DeVorkin, David H. "Michelson and the Problem of Stellar
        Diameters." *Journal for the History of
        Astronomy*, 6 (1975): 1-18.

    Discusses the use of interferometer observations
to measure stellar diameters, first attempted by Michelson
in 1890, and repeated successfully in 1920.

384. Eisele, Carolyn. "Charles S. Peirce: Nineteenth Century
        Man of Science." *Scripta Mathematica*, 24
        (1959): 305-324.

    Surveys Peirce's career and accomplishments,
focusing on his activities as an astronomer and geodesist.

385. Goldfarb, Stephen. "Science and Democracy: A History
        of the Cincinnati Observatory, 1842-1872."
        Ohio History, 78 (1969): 172-178, 222-223.

    Presents the history of the observatory as an
exemplar of the interaction of science and democracy.
Claims the observatory failed to make important contributions to astronomy because of financial difficulties
and the incompatibility of democratic use and research.

386. Gorman, Mel. "Gassendi in America." *Isis*, 55 (1964):
        409-417.

    Argues that Gassendi's 1647 textbook assisted
in the erosion of Aristotelianism and paved the way
for the acceptance of Coperican science.

387. Greene, John C. "Some Aspects of American Astronomy,
        1750-1815." *Isis*, 45 (1954): 339-358.

    Finds that Americans made no major contributions
to astronomy during this period but credits them with
keeping abreast of the latest European developments,
making and publishing useful observations, and developing
theoretical explanations for astronomical phenomena.
Summarizes American activities regarding comets, meteors,
transits, eclipses, and the dissemination of new
theoretical concepts.

388. Grosser, Morton. "The Search for a Planet Beyond
        Neptune." *Isis*, 55 (1964): 163-183.

    Describes the efforts to explain anomalies in
the motion of Neptune through predictions of a ninth
planet and the subsequent attempts to discover that
planet.

389. Hart, R., and R. Berendzen. "Hubble's Classification of
        Non-Galactic Nebulae, 1922-1926." *Journal for
        the History of Astronomy*, 2 (1971): 109-119.

    Examines the evolution of the standard
classification scheme from initial formulation to
publication.

390. Herrmann, D.B. "B.A. Gould and his *Astronomical Journal*."
        Journal for the History of Astronomy, 2 (1971):
        98-108.

    Presents an example of a German scientific
institution serving as a model for its American counterpart: in this case, a specialized journal copied its
form and objectives from the *Astronomische Nachrichten*.

391. Hetherington, Norriss S. "Cleveland Abbe and a View of
        Science in Mid-Nineteenth-Century America."
        Annals of Science, 33 (1976): 31-49.

    Discusses Abbe's efforts between 1856 and
1871 to establish a professional astronomical career.
Concludes that although there were some opportunities in

educational, research, and governmental institutions, the barriers to a professional scientific career were still greater than in Europe.

392. Hindle, Brooke. *David Rittenhouse*. Princeton: Princeton University Press, 1964. Reprint. New York: Arno Press, 1980. ix + 394 pp. Bibliography, Index.

Discusses the psychological pressures on Rittenhouse to meet the expectations of his community as he evolved from skilled craftsman to living symbol and myth. Illuminates the relationship between scientific reputation and political influence. Demonstrates the negative impact public commitments and civic responsibilities can have upon scientific activities.

393. Hoskin, M.A. "The 'Great Debate': What Really Happened." *Journal for the History of Astronomy*, 7 (1976): 169-182.

Rejects the view that the published papers represent a verbatim record of the confrontation between Heber D. Curtis and Harlow Shapley over the dimensions of the Milky Way and the status of spiral nebulae.

394. ———. "Ritchey, Curtis and the Discovery of Novae in Spiral Nebulae." *Journal for the History of Astronomy*, 7 (1976): 47-53.

Argues that it was Curtis's continuing involvement, not Ritchey's fortuitous discovery, which marks the true pioneering effort in the study of variable stars in spirals. Calls for a revision of the received account.

395. Hoyt, William Graves. *Lowell and Mars*. Tucson: University of Arizona Press, 1976. xv + 376 pp. Bibliography, Index.

Examines the interwoven history of Percival Lowell, Lowell Observatory, and research on and speculation about Mars at the turn of the twentieth century. Credits Lowell with transforming questions about the nature of Mars from scientific issues to "a raging public controversy" often aired in the popular press.

396. ————. *Planets X and Pluto*. Tucson: University of
     Arizona Press, 1980. xiv + 302 pp.
     Bibliography, Index.

     Documents the efforts of the Lowell Observatory
to locate a trans-Neptunian planet, a search capped by
the discovery of Pluto. Places the search within the
contexts of the history of planetary discovery and the
research program of the observatory. Analyzes the
factors which influenced the pace of the search--finances,
personnel, competition, and the interests of Percival
Lowell, founder and director of the observatory. Assesses
the role of improvements in instrumentation in the
discovery. Evaluates the subsequent controversy over
the identification of Pluto with Lowell's predicted
Planet X beyond Neptune.

397. ————. "W.H. Pickering's Planetary Predictions and
     the Discovery of Pluto." *Isis*, 67 (1976):
     551-564.

     Chronicles the efforts of the most prolific
postulator of undiscovered planets.

398. Inkster, Ian. "Robert Goodacre's Astronomy Lectures
     (1823-1825), and the Structure of Scientific
     Culture in Philadelphia." *Annals of Science*,
     35 (1978): 353-363.

     Contends that the success of this English
schoolmaster's lectures was due to his skill on the
platform and his arrival at a time when the scientific
institutions of Philadelphia were unprepared to meet
the growing demands for popular, inexpensive scientific
lectures.

399. Jones, Bessie Zaban. *Lighthouse of the Skies: The
     Smithsonian Observatory: Background and
     History, 1846-1955*. Washington, D.C.:
     Smithsonian Institution Press, 1965.
     xv + 339 pp. Appendices, Index.

     Provides a history of the Smithsonian prior
to the establishment of its observatory in 1892. Surveys
the physical growth, scientific achievements, and
administrative developments of the observatory.

400. ————, and Lyle G. Boyd. *The Harvard College Observatory:*
            The First Four Directorships, 1839-1919.
            Cambridge, Mass.: Harvard University Press,
            1971. xiv + 495 pp. Index.

        Chronicles both the physical development and
the scientific accomplishments of the leading American
research observatory during the nineteenth century.
Explores in depth the problems of raising an endowment
for the support of research. Contains a detailed account
of the utilization of female assistants, a pioneer effort
in the large-scale employment of women in scientific
positions.

401. Love, J. Barry. "The Miniature Solar Systems of David
            Rittenhouse." *The Smithsonian Journal of*
            History, 3, No. 4 (1968-69): 1-16.

        Discusses in detail the planetariums and
orreries constructed by David Rittenhouse. Points out
that these mechanisms were used both as demonstration
pieces for astronomy and as illustrations of the
stability of the solar system (and hence the orderli-
ness and stability of the Newtonian world system).

402. McCormmach, Russell. "Ormsby MacKnight Mitchel's
            Sidereal Messenger, 1846-1848." *Proceedings*
            of the American Philosophical Society, 110
            (1966): 35-47.

        Describes the history of the first American,
specialized, popular astronomical journal. Argues that
the journal represented an attempt to strengthen the
position of astronomy in the United States by combining
professional commitment with popular enthusiasm.

403. Musto, David F. "A Survey of the American Observatory
            Movement, 1800-1850." *Vistas in Astronomy*,
            9 (1968): 87-92.

        Analyzes the rapid increase in the number of
American observatories after 1830, concentrating on the
diversity of financial arrangements. Contrasts these
successes to the failures of the first three decades
of the century.

404. Norberg, Arthur L. "Simon Newcomb's Early Astronomical
            Career." *Isis*, 69 (1978): 209-225.

        Divides Newcomb's formative years (1857-1870)
into an "educational-calculation phase," when he acquired

and then utilized the skills of a mathematical astronomer, and an "observational phase," when he gained experience in that aspect of astronomy.

405. Obendorf, Donald Leroy. "Samuel P. Langley: Solar Scientist, 1867-1891." Ph.D. dissertation, University of California at Berkeley, 1969.

   Discusses Langley's career prior to his preoccupation with aerodynamics, concentrating on his work at the Allegheny Observatory. Focuses on his efforts at time regulation and his research on radiant energy. Argues that the driving force in Langley's life was his belief that scientific research would ultimately benefit mankind; science was done for its utility, not for personal pleasure.

406. Olson, Richard G. "The Gould Controversy at Dudley Observatory: Public and Professional Values in Conflict." *Annals of Science*, 27 (1971): 265-276.

   Presents this mid-nineteenth-century confrontation between professional scientists and their lay patrons in terms of differing visions of science and the question of deference. Was science an elite activity or should it be presented in a format accessible to all who were interested? Should the scientists be granted deference in acknowledgement of their superior knowledge or should lay trustees exert their legal authority in spite of conflicts between their desires and the wishes of the scientists?

407. Plotkin, Howard. "Edward C. Pickering, the Henry Draper Memorial, and the Beginnings of Astrophysics in America." *Annals of Science*, 35 (1978): 365-377.

   Describes the events leading to, and the results from, the Harvard College Observatory program of photographing and classifying spectra funded by Mrs. Henry Draper. The resulting catalog was one of the most important publications in early twentieth-century astrophysics.

408. ———. "Henry Draper, the Discovery of Oxygen in the
             Sun, and the Dilemma of Interpretating the
             Solar Spectrum." *Journal for the History of
             Astronomy*, 8 (1977): 44-51.

       Maintains that English astronomers did not
  objectively respond to Draper's claim of oxygen in the
  solar spectrum because of professional jealousy and
  the infusion of political disputes within the Royal
  Astronomical Society into scientific issues.

409. ———. "Henry Tappan, Franz Brünnow and the Founding
             of the Ann Arbor School of Astronomers, 1852-
             1863." *Annals of Science*, 37 (1980): 287-302.

       Describes the successful efforts of Tappan,
  the first president of the University of Michigan, and
  Brünnow, the first director of its observatory, to make
  the University an important center for the training of
  astronomers.

410. Pursell, Carroll W., Jr. *Astronomy in America*. Chicago:
             Rand McNally and Company, 1967. 48 pp.
             Chronology.

       Presents an overview of the institutional
  developments. Lacks an adequate discussion of scientific
  accomplishments.

411. Reingold, Nathan. "Cleveland Abbe at Pulkowa: Theory
             and Practice in the Nineteenth Century Physical
             Sciences." *Archives internationales d'histoire
             des sciences*, 17 (1964): 133-147.

       Places Abbe's career and work within the context
  of the discipline of geography, a field in which participants were equally at home in basic research or practical
  application.

412. Rothenberg, Marc. "The Educational and Intellectual
             Background of American Astronomers, 1825-1875."
             Ph.D. dissertation, Bryn Mawr College, 1974.

       Finds that eighty percent of American astronomers
  received some form of formal astronomical education.
  Argues that there was a strong correlation between the
  quality of the education and the quality of the subsequent
  astronomical research.

413. Warner, Deborah Jean. "Astronomy in Antebellum America."
      *The Sciences in the American Context: New
      Perspectives* (item 84), pp. 55-75.

      Credits John Herschel's *Treatise on Astronomy*
   (1834) with triggering a phenomenal growth in astronomical
   activity by making Americans aware of the interesting
   questions modern astronomy could answer. Claims astrophysics was the branch of astronomy in which Americans
   made their most important contributions during this era.

414. Webb, George Ernest. "The Scientific Career of A.E.
      Douglass, 1894-1962." Ph.D. dissertation,
      University of Arizona, 1978.

      Discusses Douglass's contributions to astronomy
   as a member of the staff of Lowell Observatory, a faculty
   member of the University of Arizona, and director of the
   Steward Observatory. Credits Douglass as the founder of
   dendrochronology. Characterizes his approach to science
   as a dedication to data collecting and individual,
   rather than team, research.

415. Wright, Helen. *Explorer of the Universe: A Biography
      of George Ellery Hale*. New York: E.P. Dutton
      & Company, 1966. 480 pp. Bibliography, Index.

      Presents an authoritative chronicle of the
   personal life and professional accomplishments of the
   central figure in the institutional development of
   American astrophysics. Balances accounts of Hale's
   scientific and technological achievements with discussions
   of his efforts to raise funds and reform the organizational structure of American science.

416. ————. *Sweeper in the Sky: The Life of Maria Mitchell,
      First Woman Astronomer in America*. New York:
      The Macmillan Company, 1949. vii + 253 pp.
      Bibliography, Index.

      Provides a study of a scientist, Vassar pro-
   fessor, and symbol of female achievement. Demonstrates
   how personal and family relationships could at least
   partially overcome the nineteenth-century obstacles to
   scientific careers for women. Discusses her activities
   as a leader of the American Association for the
   Advancement of Women.

417. Yeomans, Donald K. "The Origin of North American
               Astronomy--Seventeenth Century." *Isis*, 68
               (1977): 414-425.

        Surveys the astronomical activities of the
     English colonists.

CHEMISTRY

418. Beardsley, Edward H. *The Rise of the American Chemical
               Profession, 1850-1900*. University of Florida
               Monographs in Social Science, No. 23.
               Gainesville: University of Florida Press,
               1964. iii + 76 pp. Bibliography.

        Examines the institutional development of
     American chemistry, focusing on education, national
     professional societies, national journals, and employment
     patterns. Argues that it was the structure established
     in the second half of the nineteenth century which
     enabled twentieth-century chemists to be so successful.

419. Browne, Charles Albert, and Mary Elvira Weeks. *A History
               of the American Chemical Society: Seventy-five
               Eventful Years*. Washington, D.C.: American
               Chemical Society, 1952. xi + 526 pp. Index.

        Concentrates on the period after 1926. Supplies
     detailed descriptions of the anniversary celebrations.
     Provides a history of the publications of the ACS.
     This official history is short on analysis, but is an
     excellent source of factual material.

420. Cohen, I. Bernard. "The Beginning of Chemical Instruction
               in America: A Brief Account of the Teaching
               of Chemistry at Harvard Prior to 1800."
               Chymia, 3 (1950): 17-44.

        Rejects the thesis that the teaching of chemistry
     at Harvard originated in the Medical School in the late
     eighteenth century. Shows that chemistry has been part
     of the Harvard (and hence, American) college curriculum
     since 1687. Describes the information made available
     to Harvard students.

421. Hannaway, Owen. "The German Model of Chemical Education
               in America: Ira Remsen at Johns Hopkins."
               Ambix, 23 (1976): 145-164.

        Discusses Remsen's self-conscious effort to
     model his department upon his image of chemical education

at German universities, and the modifications imposed by the American context. Concludes that Remsen saw his role as the trainer of college teachers who in turn would educate good citizens.

422. Hornix, W.J. "The Thermostatics of J. Willard Gibbs and 19th Century Physical Chemistry." *Human Implications of Scientific Advance*. Proceedings of the XVth International Congress of the History of Science. Edited by E.G. Forbes. Edinburgh: Edinburgh University Press, 1978, pp. 314-329.

Discusses Gibbs's paper "On the Equilibrium of Heterogeneous Substances" in relationship to the work of eight major European contemporaries.

423. Ihde, Aaron J. "European Tradition in Nineteenth Century American Chemistry." *Journal of Chemical Education*, 53 (1976): 741-744.

Contends that immigrant chemists and transient American students have provided a steady stream of information about European developments in chemistry.

424. Jones, Daniel P. "American Chemists and the Geneva Protocol." *Isis*, 71 (1980): 426-440.

Describes a successful lobbying effort by scientists against the 1925 Geneva Protocol prohibiting chemical warfare. Credits this opposition to the feeling of the scientific community that the protocol would be detrimental to the best interests of the nation. Argues that the professional organization of American science during the early twentieth century aided the chemists in forming pressure groups.

425. Kauffman, George B. "The Reception of Mendeleev's Ideas in the United States and Mendeleev's Correspondence with American Scientists." *Archives internationales d'histoire des sciences*, 23 (1970): 87-106.

Finds that the reception granted Mendeleev's periodic law by Americans was similar to that of Europeans: it was accepted with little controversy but with little initial appreciation of its power.

426. Kevles, Daniel. "The Physics, Mathematics and Chemistry
        Communities: A Comparative Analysis." *The
        Organization of Knowledge in Modern America,
        1860-1920.* Edited by Alexandra Oleson and
        John Voss. Baltimore and London: The Johns
        Hopkins University Press, 1979, pp. 139-172.

     Argues that as these disciplines developed they
were dominated by small, highly productive, elite groups,
but that the structure of American science made it very
difficult for these elites to truly control the disciplines. Shows changes over time in productivity,
educational sites, and employment patterns. Finds
greater similarity than differences among the disciplines.
Chemistry was the largest of the three.

427. Kohler, Robert E., Jr. "Irving Langmuir and the 'Octet'
        Theory of Valence." *Historical Studies in
        the Physical Sciences,* 4 (1974): 39-87.

     Presents a case study of how differing personal
styles and professional situations can result in differing
receptions for the same theory. Shows how the reputation
and publicity efforts of Langmuir won acceptance in
1919-1920 for a theory which, as G.N. Lewis's Theory of
Shared Bonds, had been rejected or ignored just four
years earlier.

428. ———. "The Origin of G.N. Lewis's Theory of the
        Shared Pair Bond." *Historical Studies in the
        Physical Sciences,* 3 (1971): 343-376.

     Traces the evolution of the first satisfactory
description of the chemical bond (1916). Argues that
it derived from Lewis's attempts to fit dualistic models
to his cubic atom (the latter arose as a teaching device).

429. Kopperl, Sheldon J. "T.W. Richards' Role in American
        Graduate Education in Chemistry." *Ambix,*
        23 (1976): 165-174.

     Describes briefly Richards's transformation of
Harvard around the turn of the twentieth century into a
first-rate center for research and education in physical
chemistry.

430. Kuslan, Louis I. "The Founding of the Yale School of
        Applied Chemistry." *Journal of the History
        of Medicine,* 24 (1969): 430-451.

Argues that the survival of the first post-
graduate department in an American college was due to
the financial sacrifices of its two founding professors:
Benjamin Silliman, Jr. and John Pitkin Norton.

431. Miles, Wyndham D., editor. *American Chemists and*
            Chemical Engineers. Washington, D.C.:
            American Chemical Society, 1976. xii + 544
            pp. Index.

Presents over five hundred biographies. Includes
brief secondary bibliographies. The authors of these
biographies frequently utilized manuscript sources and
personal recollections.

432. ———. "Public Lectures on Chemistry in the United
            States." *Ambix*, 15 (1968): 129-153.

Discusses both the lectures and the lecturers
during the period from the mid-eighteenth through the
late nineteenth century. Finds great variation in the
motivation, educational background, and professional
identification of the lecturers. Concludes that there
is insufficient evidence to determine if these lectures
aided the development of chemistry in this country.

433. Servos, John W. "The Knowledge Corporation: A.A. Noyes
            and Chemistry at Cal-Tech, 1915-1930." *Ambix*,
            23 (1976): 175-186.

Argues that the keys to the success of the
Gates Chemical Laboratory at the California Institute of
Technology were the support of the Carnegie Institution
of Washington, which supplied one third of the budget,
and the "corporate" mode of administration utilized by
Arthur A. Noyes, director of chemical research.

434. Siegfried, Robert. "An Attempt in the United States to
            Resolve the Differences Between the Oxygen
            and the Phlogiston Theories." *Isis*, 46
            (1955): 327-336.

Discusses Samuel L. Mitchill's compromise
chemical theory, which accepted most of Lavoisier's
nomenclature, but substituted phlogiston for hydrogen.

435. Skolnik, Herman, and Kenneth M. Reese. *A Century of Chemistry: The Role of Chemists and the American Chemical Society.* Washington, D.C.: American Chemical Society, 1976. xvii + 468 pp. Index.

Examines chemical education, professionalism, publications, the impact of the government upon chemists, intrasocietal relations and governance, and the changing attitudes of the American Chemical Society towards legislative lobbying; the largest chapter describes developments in each of the specialty divisions of the ACS. Includes a very useful collection of lists of editors, award winners, officers, meetings, and membership statistics.

436. Tarbell, D.S., Ann T. Tarbell, and R.M. Joyce. "The Students of Ira Remsen and Roger Adams." *Isis,* 71 (1980): 620-626.

Contrasts Remsen's contribution to organic chemistry through the production of effective teachers with Adams's role as a teacher of outstanding industrial chemists.

437. Taylor, Kenneth L. "Two Centuries of Chemistry." *Issues and Ideas in America.* Edited by Benjamin J. Taylor and Thurman J. White. Norman: University of Oklahoma Press, 1976, pp. 267-284.

Contends that chemistry was the most representative science in America, intimately woven into the fabric of American social and economic life. Sees the cultivation, dissemination, and application of chemical knowledge as the central pursuits of American chemists. Concentrates on the institutional role of chemistry in American culture. Avoids an analysis of the chemical ideas developed by Americans.

438. Wilson, E. Bright, and John Ross. "Physical Chemistry in Cambridge, Massachusetts." *Annual Review of Physical Chemistry,* 24 (1973): 1-27.

Surveys the accomplishments of researchers at Harvard and the Massachusetts Institute of Technology.

GEOLOGY

*   Aldrich, Michele L. "American State Geological Surveys,
            1820-1845." Cited above as item 277.

            Discusses both the practical (economic) and
    intellectual questions considered by the geologists.
    Summarizes the contributions of the surveys.

439. Alexander, Nancy. *Father of Texas Geology: Robert T.*
            Hill. Dallas: Southern Methodist University
            Press, 1976. xii + 317 pp. Index.

            Discusses the career of the discoverer of the
    Comanche Series and pioneer investigator of West Indies
    geology. Notes that his personality led to problems in
    a variety of institutional settings, including the Texas
    Geological Survey, the United States Geological Survey,
    and the University of Texas.

440. Bray, Martha Coleman. *Joseph Nicollet and his Map.*
            Memoirs of the American Philosophical Society,
            140. Philadelphia: American Philosophical
            Society, 1980. xv + 300 pp. Index.

            Examines his contributions to the geological
    understanding of the upper Midwest. Provides insights
    into the subculture of French intellectuals in the
    United States.

441. Carozzi, Albert V. "Agassiz's Influence on Geological
            Thinking in the Americas." *Archives des*
            sciences, 27 (1974): 5-38.

            Argues that his influence was relatively minor.
    Finds general indifference to or rejection of his theories
    of tropical and continental glaciation.

442. Darrah, William Culp. *Powell of the Colorado.* Princeton:
            Princeton University Press, 1951. Reprint.
            xx + 426 pp. Bibliography, Index.

            Provides detailed discussions of Powell's
    contributions as explorer, founder and first director
    of the Bureau of American Ethnology, and second director
    of the United States Geological Survey. Presents Powell
    as a popularist defending natural resources against
    exploitation by monopolies. Sees him as a generally
    successful participant in the political wars in
    Washington.

443. Dean, Dennis R. "The Influence of Geology on American
            Literature and Thought." *Two Hundred Years of
            Geology in America: Proceedings of the New
            Hampshire Bicentennial Conference on the
            History of Geology* (item 461), pp. 289-303.

        Views literature as an indicator of public
    attitudes towards and knowledge of geology. Includes as
    examples the utilitarian slant of the colonists, the
    uniformitarianism assumed in the works of Washington
    Irving, and the late nineteenth-century attitude that
    man was powerless before the forces of nature.

444. Dott, R.H., Jr. "The Geosyncline--First Major Geological
            Concept 'Made in America.'" *Two Hundred Years
            of Geology in America: Proceedings of the New
            Hampshire Bicentennial Conference on the
            History of Geology* (item 461), pp. 239-264.

        Credits James Hall's extensive field experience
    with giving him the insights which led to his conception
    of the continued accretion theory of mountain building,
    later named and nurtured by James Dwight Dana.

445. Frankel, Henry. "Why the Drift Theory was Accepted with
            the Confirmation of Harry Hess's Concept of
            Sea-Floor Spreading." *Two Hundred Years of
            Geology in America: Proceedings of the New
            Hampshire Bicentennial Conference on the History
            of Geology* (item 461), pp. 337-353.

        Views the reluctance of the scientific community
    to accept Drift Theory until after the confirmation of
    the Vine-Matthews and Wilson Hypotheses--corollaries of
    Hess's concept of sea-floor spreading--as evidence of
    the hesitation of scientists to embrace a new theory
    which merely explains the known facts. Argues that
    scientists demand that a theory also predict new
    phenomena.

446. Gerstner, Patsy A. "A Dynamic Theory of Mountain Building:
            Henry Darwin Rogers, 1842." *Isis*, 66 (1975):
            26-37.

        Examines Rogers's theory that mountains were
    the result of surface upheavals caused by the wave-like
    motion of the molten matter below the crust. Concludes
    that it was generally rejected because it contradicted
    the arguments of Charles Lyell.

447. ———. "Henry Darwin Rogers and William Barton Rogers
on the Nomenclature of the American Paleozoic
Rocks. *Two Hundred Years of Geology in America:
Proceedings of the New Hampshire Conference on
the History of Geology* (item 461), pp. 175-186.

Illustrates the conceptual difficulties involved
in formulating nomenclature. Follows the evolution of
this nomenclature from a numerical format to a complex
series of names designating particular time periods.

448. Greene, John C., and John G. Burke. *The Science of
Minerals in the Age of Jefferson.* Transactions
of the American Philosophical Society, 68,
part 4. Philadelphia: American Philosophical
Society, 1978. 113 pp. Index.

Traces the development of mineralogy in the
United States through 1822, the period when the pioneering
generation of American mineralogists established a
network of researchers. Stresses the regional nature
of American activity in contrast to the centralized
European communities. Spotlights the contributions of
Parker Cleaveland, whose textbook (the first American
text on this subject) was both the epitome of European
mineralogy and an evaluation of the mineral resources
of the United States. Concludes that at the close of
this period American mineralogy was still dependent upon
and derivative of its European counterparts.

* Hendrickson, Walter B. "Nineteenth-Century State
Geological Surveys: Early Government Support
of Science." Cited above as item 308.

Finds that scientists argued that the state
should fund research because of possible economic return;
in exchange for his services to the state the scientist
received the opportunity of conducting his personal
research at the state's expense.

449. Jordan, William M. "Geology and the Industrial-Trans-
portation Revolution in Early to Mid-Nineteenth-
Century Pennsylvania. *Two Hundred Years of
Geology in America: Proceedings of the New
Hampshire Bicentennial Conference on the
History of Geology* (item 461), pp. 91-103.

Argues that the exploitation of the coal mines
and the development of the transportation system were
great stimuli to the progress of geology in Pennsylvania.

Correlates the pace of geological research to the state's economic conditions.

450. Kitts, David B. "Grove Karl Gilbert and the Concept of 'Hypothesis' in Late Nineteenth-Century Geology." *Foundations of Scientific Method: The Nineteenth Century*. Edited by Ronald N. Giere and Richard S. Westfall. Bloomington and London: Indiana University Press, 1973, pp. 259-274.

Argues that Gilbert's concept of the hypothesis as the proposal of the particular antecedent conditions has been a central thread in the methodology of American geologists. Views Gilbert as outside any familiar philosophical traditions.

451. Kohlstedt, Sally Gregory. "The Geologists' Model for National Science, 1840-1847." *Proceedings of the American Philosophical Society*, 118 (1974): 179-195.

Traces the history of the Association of American Geologists, originally established to provide uniform classification schemes and nomenclature for state geological surveys, which became the forerunner of the American Association for the Advancement of Science.

452. Lintner, Stephen F., and Darwin H. Stapleton. "Geological Theory and Practice in the Career of Benjamin Henry Latrobe." *Two Hundred Years of Geology in America: Proceedings of the New Hampshire Bicentennial Conference on the History of Geology* (item 461), pp. 107-119.

Demonstrates how Latrobe applied his knowledge of geology to his efforts in architecture and civil engineering.

453. Merrill, G.P. *The First Hundred Years of American Geology*. New Haven: Yale University Press, 1924. Reprint. New York: Hafner, 1964. xxi + 773 pp. Appendix, Index.

Divides this period into three phases: the Macluren Era, 1785-1819, centered upon the publication of William Maclure's *Observations on the Geology of the United States* (1809); the Eatonian Era, 1820-1829, dominated by Amos Eaton's *Index* and his survey of the Erie Canal route; and the period of surveys, 1830-1880, marked by a tremendous expansion of the public funding

for geology. Discusses the prevalent theoretical
positions and important field work in each of the phases.

454. Nelson, Clifford M. "William Henry Holmes: Beginning a
Career in Art and Science." *Records of the
Columbia Historical Society of Washington,
D.C.*, 50 (1980): 252-278.

Describes the role of the Smithsonian Institution and the Hayden Survey in the development of Holmes
into a geological artist, geologist, and archaeologist.

455. Parry, Elwood C., III. "Acts of God, Acts of Man:
Geological Ideas and the Imaginary Landscapes
of Thomas Cole." *Two Hundred Years of Geology
in America: Proceedings of the New Hampshire
Bicentennial Conference on the History of
Geology* (item 461), pp. 53-71.

Argues that changes in the landscape renditions
of the first Romantic landscape painter in the United
States reflected a shift in his geological orientation
from catastrophism to uniformitarianism.

456. Pyne, Stephen J. "From the Grand Canyon to the Marianas
Trench: The Earth Sciences After Darwin."
*The Sciences in the American Context: New
Perspectives* (item 84), pp. 165-192.

Asserts that during the century 1860-1960
geology has been transformed in four significant ways:
a tremendous growth in the quantity and change in the
quality of geological information; a shift in the basic
working concept from the contracting earth to plate
theory; a replacement of evolutionary views of the earth
by the conception of the earth as a recycling open system;
and the elimination of historicism.

457. ————. *Grove Karl Gilbert: A Great Engine of Research*.
Austin and London: University of Texas Press,
1980. xiv + 306 pp. Bibliography, Index.

Notes that Gilbert's lifetime (1843-1918)
spanned the period when American geology was intellectually
and institutionally defined. Argues that Gilbert was
atypical of the American geological community: he was
not fascinated with the organic world; he utilized
mechanical rather than organic metaphors; he saw nature's
pattern in terms of equilibrium rather than evolution;
he never identified himself as a teacher or leader of a

geological school. Characterizes Gilbert's approach to
geology as the application of engineering techniques to
geological subjects.

458. ————. "Methodologies for Geology: G.K. Gilbert and
T.C. Chamberlin." *Isis*, 69 (1978): 413-424.

Argues that the difference in methodology between the two major figures in the "Heroic Age" of
American geology reflected differences in their tempera-
ment, personality, and careers. Contrasts Chamberlin's
efforts to amplify the merits of competing hypotheses
until one appeared clearly superior with Gilbert's attempts
to uncover the liabilities in hypotheses until only one
explanation survived.

459. Schneer, Cecil J. "Ebenezer Emmons and the Foundations
of American Geology." *Isis*, 60 (1969): 439-450.

Summarizes Emmons's contributions to the New
York State Natural History Survey, focusing on his use
of non-Wernerian nomenclature adopted by the other New
York State geologists.

460. ————. "The Great Taconic Controversy." *Isis*, 69
(1978): 173-191.

Traces the controversy which followed Ebenezer
Emmons's attempts to demonstrate the existence of a new
system of strata representing the oldest system of
stratified rocks, a controversy which lasted through
the remainder of the nineteenth century. Concludes
that personalities and institutional politics became
intertwined with the geological observations.

461. ————, editor. *Two Hundred Years of Geology in America:
Proceedings of the New Hampshire Bicentennial
Conference on the History of Geology*. Hanover:
University Press of New England, 1979. xvi +
385 pp. Appendix, Index.

Contains twenty-six papers, some too short to
successfully develop their arguments. There are four
major historiographic approaches taken by the authors:
internal studies, focusing on specific theoretical
concepts which either arose or developed in the United
States; biographical studies; histories of the institutions which supported geological research; studies of
the interaction of geology with other aspects of the
culture. Includes items 156, 277, 443, 444, 445, 447,
449, 452, 455, 480, 602.

462. Schultz, Susan F. "Thomas C. Chamberlin: An Intellectual
              Biography of a Geologist and Educator." Ph.D.
              dissertation, University of Wisconsin-Madison,
              1976.

       Studies Chamberlin's scientific ideas within
their professional and intellectual contexts but does
not provide a discussion of his personal life. Focuses
on his work on glaciers, the origin of the earth, and
the proper scientific methodology. Argues that Chamberlin's scientific and social outlooks were explicitly
moralistic.

463. Shrock, Robert Rakes. *Geology at M.I.T., 1865-1965: A*
              History of the First Hundred Years of Geology
              at Massachusetts Institute of Technology.
              Volume I: *The Faculty and Supporting Staff.*
              Cambridge, Mass. and London: The M.I.T. Press,
              1977. xxiv + 1,032 pp.

       Studies the staff of the institution as
individuals rather than part of a dynamic system of
interacting components from both within and without the
department of geology. Includes the biographies and
bibliographies of the fifty-two geologists who were
professors at M.I.T. These are very uneven in length,
reflecting the degree to which a scientists' career was
interwoven with M.I.T., but the more complete biographies
and bibliographies are excellent. To be followed by a
volume focusing on the students and the research projects.

464. Stegner, Wallace. *Beyond the Hundredth Meridian: John*
              Wesley Powell and the Second Opening of the
              West. Boston: Houghton Mifflin Company, 1954.
              xxiii + 438 pp. Index.

       Sees Powell's objective as the imposition of
order through science. Presents Powell as the personification of public service.

465. White, George W. "Early Geological Observations in the
              American Midwest." *Towards a History of*
              Geology. Edited by Cecil J. Schneer. Cambridge,
              Mass. and London: The M.I.T. Press, 1969,
              pp. 415-425.

       Characterizes colonial geology in this region
as observational rather than analytical.

466. ———. "The History of Geology and Mineralogy as Seen
         by American Writers, 1803-1835: A Bibliographic
         Essay." *Isis*, 64 (1973): 197-214.

         Summarizes seven attempts to write a history of
    geology by Americans of differing intellectual backgrounds.
    Finds that the histories reflect both the intellectual
    matrix of the writer and changing American attitudes
    towards geological theories.

467. Wilkins, Thurman. *Clarence King: A Biography*. New York:
         The Macmillan Company, 1958. ix + 441 pp.
         Bibliography, Index.

         Relates the life of the head of the Fortieth
    Parallel Survey, first director of the United States
    Geological Survey, and member of the genteel, literary
    society of the Gilded Age. Credits King with the princi-
    pal role in the establishment of the Geological Survey.

468. Wilson, Leonard G., editor. *Benjamin Silliman and His
         Circle: Studies on the Influence of Benjamin
         Silliman on Science in America--Prepared in
         Honor of Elizabeth H. Thomson*. New York:
         Science History Publications, 1979. x +
         228 pp. Bibliography, Index.

         Focuses on Silliman's roles as teacher and
    lecturer.

469. ———. "The Emergence of Geology as a Science in the
         United States." *Journal of World History*, 10
         (1967): 416-437.

         Argues that the United States had a group of
    competent professional geologists doing work of inter-
    national significance by the mid-nineteenth century.
    Claims that this rapid development was made possible
    by the abundance of material available for study, the
    economic expansion which funded geological research, and
    the rapid assimilation of European theoretical advances.

470. Winnik, Herbert C. "Science and Morality in Thomas C.
         Chamberlin." *Journal of the History of Ideas*,
         31 (1970): 441-456.

         Argues that major themes in Chamberlin's science,
    including a search for fundamentals, a belief in the
    harmony of nature, and an optimistic, progressive
    evolutionary theory, were the result of his Methodist
    upbringing and his father's concern for social issues.

MATHEMATICS

471. Birkhoff, Garrett. "Some Leaders in American Mathematics, 1891-1941." *The Bicentennial Tribute to American Mathematics, 1776-1976*. Edited by Dalton Tarwater. N.P.: The Mathematical Association of America, 1977, pp. 25-78.

Surveys the great deeds and men during the period from the development of a mature mathematical community to the obtainment of world leadership.

472. Grabiner, Judith V. "Mathematics in America: The First Hundred Years." *The Bicentennial Tribute to American Mathematics, 1776-1976*. Edited by Dalton Tarwater. N.P.: The Mathematical Association of America, 1977, pp. 9-24.

Credits the creation of a world-class mathematics discipline in the United States between 1876 and 1900 to the slow but steady upgrading of the quality of American mathematics education during the last half of the nineteenth century.

473. Hamming, R.W. "The History of Computers in the United States." *The Bicentennial Tribute to American Mathematics, 1776-1976*. Edited by Dalton Tarwater. N.P.: The Mathematical Association of America, 1977, pp. 101-116.

Calls for more studies of software. Contends that most mathematicians have ignored computers because the return, at least initially, did not meet their expectations.

474. Heims, Steve J. *John von Neumann and Norbert Wiener: From Mathematics to the Technologies of Life and Death*. Cambridge, Mass. and London: The M.I.T. Press, 1980. xviii + 547 pp. Index.

Presents parallel biographies of two scientists, contrasting their responses to the atomic bomb. Finds that both men shifted their research from pure science to the service of explicit social and political values, but that the values differed. Contrasts von Neumann's postwar decision to remain in the corridors of power, assisting the military in weapons research, with Wiener's refusal to continue in weapons research, choosing instead to worry about the negative social and moral implications of scientific research. The author is explicitly against

weapons research. He is also simplistic, if not
inaccurate, in his efforts to interpret historial events.

475. Hogan, Edward R. "Robert Adrain: American Mathematician."
     Historia Mathematica, 4 (1977): 157-172.

          Summarizes the life and contributions of one
     of the first American mathematicians to publish original
     research. Views Adrain as a transitional figure during
     the period of the assimilation of French mathematics and
     increasing concern with research by the American community.

476. Hughes, Thomas Parke. "ENIAC: Invention of a Computer."
     Technikgeschichte, 42 (1975): 148-165.

          Argues that the invention of this computer was
     not a discrete event, but rather, like most inventions,
     the culmination of a number of innovations. Sees the
     computer as a response to a reverse salient in military
     ordnance technology. Identifies the critical problem
     as a reliable counter.

*   Kevles, Daniel. "The Physics, Mathematics and Chemistry
                Communities: A Comparative Analysis." Cited
                above as item 426.

          Finds that the mathematics community was the
     smallest of the three but was the best developed. Shows
     that American mathematicians were much more likely to
     publish their results in foreign journals than their
     counterparts in the other two disciplines.

477. Pycior, Helena M. "Benjamin Peirce's *Linear Associative
                Algebra*." *Isis*, 70 (1979): 537-551.

          Examines the background and creation of a
     major contribution to the development of algebra.
     Advocates the thesis that the book was generally ignored
     by contemporaries because it was viewed as a philosophical
     rather than mathematical work. Argues that Peirce
     attempted to legitimatize his algebra through a theological
     appeal.

METEOROLOGY AND OCEANOGRAPHY

478. Burstyn, Harold L. *At the Sign of the Quadrant: An
                Account of the Contributions to American
                Hydrography Made by Edmund March Blunt and
                his Sons*. Mystic: The Marine Historical
                Association, 1957. 119 pp. Bibliography.

Argues that the Blunts filled the need created by the refusal of the American government to supply navigation charts, a situation which remained in effect until 1867. Provides biographical information on E.M. Blunt and his sons, Edmund and George W., who succeeded their father in the business in 1833. Demonstrates the links between private enterprise and public service through the career of George W. Blunt, who was both First Assistant with the Coast Survey and a member of the family business.

479. ————. "Reviving American Oceanography: Frank Lillie, Wickliffe Rose, and the Founding of the Woods Hole Oceanographic Institution." *Oceanography: The Past*. Edited by M. Sears and D. Merriman. Proceedings of the Third International Congress on the History of Oceanography. New York, Heidelberg, and Berlin: Springer-Verlag, 1980, pp. 57-66.

Describes the efforts of the chairman of the National Academy of Science's Committee on Oceanography (Lillie) and the President of the General Education Board of the Rockefeller Foundation (Rose) to restore American oceanography to a position at the forefront of research, a position it had held in the mid-nineteenth century.

480. ———— and Susan B. Schlee. "The Study of Ocean Currents in America Before 1930." *Two Hundred Years of Geology in America: Proceedings of the New Hampshire Bicentennial Conference on the History of Geology* (item 461), pp. 145-155.

Blames the post Civil War shift in interest from oceans to the Western interior for the decline of American physical oceanography from its earlier dominant position.

481. Haedrich, Richard L., and Kenneth O. Emery. "Growth of an Oceanographic Institution." *Oceanography: The Past*. Edited by M. Sears and D. Merriman. Proceedings of the Third International Congress on the History of Oceanography. New York, Heidelberg, and Berlin: Springer-Verlag, 1980, pp. 67-82.

Argues that the development of Woods Hole since 1930 has been a microcosm of the growth of American science generally. Emphasizes the role of military funding, the impact of the National Science Foundation, and the re-

evaluations forced by tremendous inflationary pressures
upon budgets. Contends that one result of the increase
in budget levels was the growth of specialization
within oceanography.

482. Leighly, John. "Introduction." *The Physical Geography
          of the Sea and its Meteorology*. By Matthew
          Fontaine Maury. Cambridge, Mass.: Harvard
          University Press, 1963, pp. ix-xxx.

     Demonstrates the errors, deficiencies, and
invalid hypotheses which abound in Maury's effort to
apply physical theory to the atmosphere and oceans in
order to derive an explanation of the earth's climate.
Argues that the book did not come up to the standards
of the physical geography of its day. Suggests that its
popularity was a result of its literary qualities.

483. Ludlum, David M. *Early American Tornadoes, 1586-1870*.
          History of American Weather. Boston: American
          Meteorological Society, 1970. vii + 219 pp.
          Indices.

     Combines a chronological and geographical
accounting of all recorded tornadoes with a description
of American attempts to analyze and understand such
storms. Focuses on the work of James P. Espy and
William C. Redfield.

484. Raitt, Helen, and Beatrice Moulton. *Scripps Institution
          of Oceanography: First Fifty Years*. Los
          Angeles: The Ward Ritchie Press, 1967. xix +
          205 pp. Appendices, Bibliography, Index.

     Traces the history of Scripps through World War
II as it evolved from a small, private marine biology
research group to an independent, endowed division of
the University of California. Observes that oceanography
was not emphasized at Scripps until after World War I,
and even as late as the 1930s, most Scripps's scientists
were not involved in research problems which required
deep water facilities. Credits Scripps with developing
the first American degree program in oceanography.

485. Schlee, Susan. *On Almost Any Wind: The Saga of the
          Oceanographic Research Vessel* Atlantis. Ithaca
          and London: Cornell University Press, 1978.
          301 pp. Index.

     Studies the activities of the Woods Hole
Oceanographic Institute's first deep-water research

vessel, one of the very few such vessels in America during
the years 1931-1966. Provides insights on life aboard
such a ship. Assesses the scientific contributions
made possible by its use.

486. Shor, Elizabeth Noble. *Scripps Institution of Oceanogra-
           phy: Probing the Oceans, 1936-1976.* San Diego:
           Tofua Press, 1978. x + 502 pp. Appendices,
           Index.

     Updates item 484. Divides the presentation
into a chronological history through the end of World
War II, a discussion of the various research units, an
analysis of the contributions of Scripps to a variety of
disciplines, and a survey of the expeditions and facilities. The author frequently loses the thread of her
argument in a complex of side issues and interesting but
irrelevant anecdotes.

487. Sinclair, Bruce. "Gustavus A. Hyde, Professor Espy's
           Volunteers and the Development of Systematic
           Weather Observations." *Bulletin of the
           American Meteorological Society*, 46 (1965):
           779-784.

     Traces the evolution of systematic weather
observations during the last half of the nineteenth
century through a brief study of the career of a volunteer
data collector.

488. Spence, Clark C. *The Rainmakers: American "Pluviculture"
           to World War II.* Lincoln and London: University
           of Nebraska Press, 1980. x + 181 pp. Index.

     Views most rainmakers as confidencemen exploit-
ing the needs and fears of farmers and businessmen in
farming communities. Divides rainmaking techniques into
five categories: upward convection of warm air, explosives,
chemicals, electricity, and cloud-salting from the air.
Shows that almost all professional meteorologists rejected rainmaking ideas, but that the press usually
provided the rainmakers with favorable coverage.

489. Williams, Frances Leigh. *Matthew Fontaine Maury:
           Scientist of the Sea.* New Brunswick: Rutgers
           University Press, 1963. xx + 720 pp.
           Bibliography, Index.

     Presents a partisan account of Maury's struggles
with A.D. Bache, Joseph Henry, and their allies for the

leadership of the American physical science community.
Describes Maury's contributions to astronomy, meteorology,
and oceanography, his administration of the Naval Observatory, and his efforts to improve the training of naval
officers.

490. Zinn, Donald. "Alexander Agassiz (1835-1910) and the
                Financial Support of Oceanography in the United
                States." *Oceanography: The Past*. Edited by
                M. Sears and D. Merriman. Proceedings of the
                Third International Congress on the History of
                Oceanography. New York, Heidelberg, and Berlin:
                Springer-Verlag, 1980, pp. 83-93.

        Describes Agassiz as a catalyst in oceanographic
research. Surveys his international network of connections
among both scientists and financial supporters of science.
Examines his own activities as a scientist and businessman.

PHYSICS AND GEODESY

491. Badash, Lawrence. *Radioactivity in America: Growth and
                Decay of a Science*. Baltimore and London:
                The Johns Hopkins University Press, 1979.
                xviii + 327 pp. Index.

        Argues that radioactivity was very representative
of the state of American science in general and physics
in particular circa 1900-1920. Finds that American
research in this field was limited by shortages of manpower, funds, and genius, shortages not uncommon in a
developing country. Observes that most American research
was experimental, respectable, and of interest to
Europeans but did not represent breakthroughs or notable
discoveries. Notes the superiority of the international
scientific communications network to the domestic network;
American scientists were often in closer contact with
their European counterparts than with their American
colleagues in the same specialty.

492. Berry, Ralph Moore. "History of Geodetic Leveling in
                the United States." *Surveying and Mapping*,
                36 (1976): 137-153.

        Chronicles the use of this highly accurate
technique since its introduction by the Coast Survey
in 1856.

*   Bray, Martha Coleman. *Joseph Nicollet and His Map*.
              Cited above as item 440.

         Credits this French-born scientist with intro-
ducing new methods of accurate mapping to the United
States. Appears to discount the role of West Point as
a training ground for military astronomers, claiming
that such men were actually the disciples of Nicollet.

493. Brown, Sanborn C., and Leonard M. Rieser. *Natural
              Philosophy at Dartmouth: From Surveyors' Chains
              to the Pressure of Light*. Hanover: The
              University Press of New England, 1974. xii +
              127 pp. Index.

         Describes the lives and work of the physical
scientists on the Dartmouth faculty. Demonstrates in-
,creasing specialization and commitment to research.
Ignores the background and connections with general
trends in American science.

494. Childs, Herbert. *An American Genius: The Life of Ernest
              Orlando Lawrence, Father of the Cyclotron*.
              New York: E.P. Dutton & Company, 1968.
              576 pp. Index.

         Offers a detailed, sympathetic account of the
life and scientific accomplishments of the 1939 Nobel
Prize winner in physics, who was both a leader in
experimental high energy physics and a major figure in
science administration. Draws upon interviews and
correspondence to construct the definitive chronicle
but fails to provide an analytical, interpretive biography.

495. Coben, Stanley. "The Scientific Establishment and the
              Transmission of Quantum Mechanics to the United
              States, 1919-32." *American Historical Review*,
              76 (1971): 442-466.

         Finds that the leadership of the American
physics community during the second decade of the
twentieth century was sympathetic to, and developed the
financial support for, the pursuance of training in
theoretical physics, paving the way for the rapid
assimilation of quantum theory in this country.

496. Cohen, I. Bernard. "American Physicists at War: From
        the Revolution to the World Wars." *American
        Journal of Physics*, 13 (1945): 223-235.

        States that American physicists have played an
ever-increasing role in the conduct of war. Credits this
to both the growth of American physics and the increasing
importance of science in modern civilization. Provides
examples of physical scientists contributing to the
national effort rather than concrete examples of physics
applied to warfare.

497. ———. "American Physicists at War: From the First
        World War to 1942." *American Journal of
        Physics*, 13 (1945): 333-346.

        Continues item 496. Concentrates on the efforts
of the National Research Council. No longer viewed as
reliable.

498. ———. *Franklin and Newton: An Inquiry into Speculative
        Newtonian Experimental Science and Franklin's
        Work in Electricity as an Example Thereof.*
        Memoirs of the American Philosophical Society,
        43. Cambridge: Harvard University Press, 1966.
        xxvi + 657 pp. Appendices, Bibliography, Index.

        Describes the Newton of the *Opticks* as a
speculative experimentalist, and claims that Franklin was
in the same mold. Argues that Franklin was the first to
show that electrical phenomena were important for understanding the world; he was also the first to show the
importance of the concept of conservation, thus indicating
the possibility of a completely mechanical explanation
of the universe. Provides a detailed description of
Franklin's experiments and theories, their reception in
Europe, and the evolution of electrical theory through
the end of the eighteenth century.

499. ———. "The Two Hundredth Anniversary of Benjamin
        Franklin's Two Lightning Experiments and the
        Introduction of the Lightning Rod." *Pro-
        ceedings of the American Philosophical Society*,
        96 (1952): 331-366.

        Defends Franklin against accusations that the
experiments were either historical myths or, if performed,
plagiarisms of the European sentry-box experiments.

500. Davis, Nuel Pharr. *Lawrence and Oppenheimer*. Greenwich, Conn.: Fawcett Publications, 1969. 384 pp. Appendices, Glossary, Bibliography, Index.

Compares and contrasts the personalities and professional careers of the two giants of American physics during the middle third of the twentieth century: Ernest O. Lawrence, the experimentalist who invented the cyclotron, and J. Robert Oppenheimer, the theorist who became director of Los Alamos. Depends heavily upon interviews with colleagues. Suffers from inaccuracies, but does supply a gossipy view of the physics community as it struggled to come to terms with big science, the atomic bomb, and its own sense of guilt over the destruction science could cause.

501. Finn, Bernard S. "An Appraisal of the Origins of Franklin's Electrical Theory." *Isis*, 1969 (60): 362-369.

Argues that Franklin's ignorance of scientific activity in Europe allowed him to create a theory without regard for the experimental evidence which was proving so indigestible for European theorists.

502. Heilbron, J.L. "Franklin, Haller, and Franklinist History." *Isis*, 68 (1977): 539-549.

Identifies Albrecht von Haller as Franklin's first guide in electricity. Attacks the prevalent historiography for eighteenth-century electricity as Franklinist--bias against those who rejected Franklin's views.

503. Hindle, Brooke. "Cadwallader Colden's Extension of the Newtonian Principles." *The William and Mary Quarterly*, 3rd ser., 13 (1956): 459-475.

Examines Colden's efforts to utilize Newton's *Opticks* in order to discover the cause of gravity. Concludes that Colden was a victim of faulty logic, his misunderstanding of Newton's laws of motion, and his inability to distinguish between physical and metaphysical questions.

504. Home, Roderick W. "Franklin's Electrical Atmospheres." *The British Journal for the History of Science*, 6 (1972): 131-151.

Describes the contradictions that arose through Franklin's use of electrical atmospheres to explain

attraction and repulsion. Examines some of the efforts
by later physicists to evolve a more satisfactory theory.

505. Joncich, Geraldine. "Scientists and the Schools of the
        Nineteenth Century: The Case of American
        Physicists." *American Quarterly*, 18 (1966):
        667-685.

        Examines the autobiographical accounts of
American physicists born in the nineteenth century.
Finds that they were generally dissatisfied with the
quality of their science teachers and courses on the
primary, secondary, and undergraduate levels.

506. Kargon, Robert H. "The Conservative Mode: Robert A.
        Millikan and the Twentieth-Century Revolution
        in Physics." *Isis*, 68 (1977): 509-526.

        Classifies Millikan as a conservative during a
revolutionary era in physics. Claims that his desire
for precision, evident in his measurement of the charge
upon the electron, was an attempt both to bring a sense
of stability to his world and to draw upon the American
tradition of demonstrating excellence in physics through
precise measurement.

507. Kevles, Daniel J. "On the Flaws of American Physics:
        A Social and Institutional Analysis." *Nine-
        teenth Century American Science: A Reappraisal*
        (item 32), pp. 133-151.

        Argues that the lack of national institutions
to establish standards of excellence, guide the profession, and ask important questions caused the relatively
poor performance of American physics during the late
nineteenth century.

508. ————. *The Physicists: The History of a Scientific
        Community in Modern America*. New York: Alfred
        A. Knopf, 1978. xi + 489 pp. Bibliography,
        Index.

        Examines the evolution of the discipline and
its interactions with broader scientific and cultural
developments. Focuses on the issue of patronage. Contends that the physics community held allegiance to two
forms of elitism in the distribution of funds which
contradicted the ideals of American democracy: "best
science," which led to an aristocracy of leading
researchers garnering the lion's share of the available

resources, and "political," which argued that the
scientific community, not the elected political leadership,
should determine the distribution of public funds desig-
nated for the support of research.

*   ————. "The Physics, Mathematics and Chemistry Communities: A Comparative Analysis." Cited above
as item 426.

Demonstrates the domination of the discipline
from 1870 through World War I by a few graduate departments.

509. Klein, M.J. "The Early Papers of J. Willard Gibbs: A
Transformation of Thermodynamics." *Human
Implications of Scientific Advance.* Proceedings
of the XVth International Congress of the History
of Science. Edited by E.G. Forbes. Edinburgh:
Edinburgh University Press, 1978, pp. 330-341.

Rejects the thesis that Gibbs's first two
papers--on geometrical methods in thermodynamics--were
important only as the preparation of the ground for his
later work. Finds that Gibbs transformed thermodynamics
in these papers by establishing entropy as a fundamental
concept.

510. Lenzen, Victor F. "Charles S. Peirce as Mathematical
Geodesist." *Transactions of the Charles S.
Peirce Society*, 8 (1972): 90-105.

Examines Peirce's contributions to the
determination of the ellipticity of the earth.

511. Livingston, Dorothy Michelson. *The Master of Light: A
Biography of Albert A. Michelson.* New York:
Charles Scribner's Sons, 1973. xi + 376 pp.
Chronology, Index.

Offers insights into the personality of America's
first Nobel laurate for science and the interaction of
his private life, public career, and scientific activities.
Examines Michelson's contributions to science, delineating
his strengths and weaknesses in the field of physics.

512. Miller, John David. "Rowland and the Nature of Electric
Currents." *Isis*, 63 (1972): 5-27.

Describes Henry Augustus Rowland's series of
experiments to ascertain whether electricity acted like
a fluid. Shows that Rowland practiced experimental
physics on a par with the best of the European scientists.

513. ————. "Rowland's Magnetic Analogy to Ohm's Law."
        Isis, 66 (1975): 230-241.

        Reconstructs Rowland's early research in magnet-
ism, emphasizing his utilization of models and analogies
to describe physical phenomena.

514. Moyer, Albert Earl. "Conceptual Shifts in Late Nineteenth-
        Century American Physics." Ph.D.. dissertation,
        University of Wisconsin-Madison, 1977.

        Studies the shift from a mechanical world view
to mathematically sophisticated, operational descriptions.
Finds that the mechanical world view was dominant but not
unchallenged from 1870 through 1895. Discovers considerable intellectual unrest among physical scientists
post-1895. Demonstrates the strong influence of European
ideas, theories, and concepts upon the American community.

515. Mozley, Ann. "Change in Argonne National Laboratory: A
        Case Study." *Science*, 174 (1971): 30-38.

        Finds that the science staff at Argonne resisted
changes in administration made by the Argonne University
Association. Discusses the staff's feelings that the
Association has shown little interest in long-term
policy and a tendency to overadminister the day-to-day
activities.

516. Post, Robert Charles. *Physics, Patents, and Politics:*
        A Biography of Charles Grafton Page. New York:
        Science History Publications, 1976. 227 pp.
        Bibliography, Index.

        Attempts to restore Page's reputation as a
scientist. Claims that his work in experimental physics
was on a par with his more honored contemporary Joseph
Henry but that Page's violations of the accepted code of
behavior for professional scientists--his taking out of
patents and his decision to utilize the political system
to defend his scientific reputation--resulted in distorted
evaluations of his contributions.

517. Seidel, Robert W. "The Origins of Academic Physics
        Research in California: A Study of Inter-
        disciplinary Dynamics in Institutional Growth."
        Journal of College Science Teaching, 6 (1976):
        10-23.

        Compares and contrasts the experiences of the
physics programs at the University of California (Berkeley),

California Institute of Technology, and Stanford. Finds
that the first two institutions were able to build first-
class, research-oriented departments through local
financial support and strong interdisciplinary ties to
fields like astronomy and physical chemistry. Blames
the failure of Stanford to develop an equivalent department on the lack of such support and ties.

518. Smith, Thomas M. "Physics." *Issues and Ideas in America.*
Edited by Benjamin J. Taylor and Thurman J.
White. Norman: University of Oklahoma Press,
1976, pp. 285-293.

Considers the role of the Industrial Revolution,
internal developments in European physics, and America's
·skill at organization in the evolution of physics in the
United States. Argues that the most profound change in
physics has been its entry into the political arena.

519. Sopka, Katherine Russell. *Quantum Physics in America,*
1920-1935. New York: Arno Press, 1980. xx +
540 pp. Appendices, Bibliography, Index.

Finds that the establishment of quantum theory
coincided with the maturation of the American physics
community. Divides the era into three periods: from 1920
through 1925 American physicists studied the old quantum
theory and entered in the debates over its validity; the
second period, 1926-1929, was marked by the rapid
assimilation of the revolutionary quantum mechanics by
the American community, which was very receptive to the
new trend towards abstraction; during the years 1930-
1935 quantum mechanics was confirmed, while theoretical
physics flourished in the United States. Delineates the
contributions of John H. Van Vleck and Edwin C. Kemble
in the development of theoretical physics in this country.

520. Stuewer, Roger H. *The Compton Effect: Turning Point in*
Physics. New York: Science History Publications,
1975. xii + 367 pp. Appendices, Index.

Discusses the conceptual origins and experimental
confirmation of Arthur Holley Compton's quantum interpretation for the scattering of x-rays. Examines his efforts
to provide a classical explanation, his initial rejection
of Einstein's hypothesis of the particle-like photon,
and the response to his theory (hesitation, then rapid
acceptance). This is a history of a particular scientific
concept, not a biography of Compton nor a general study
of his work.

521. ————. "G.N. Lewis on Detailed Balancing, the Symmetry
          of Time, and the Nature of Light." *Historical
          Studies in the Physical Sciences*, 6 (1975):
          469-511.

   Describes Lewis's attempt to combine Einstein's
Theory of Relativity and the light quantum hypothesis,
the symmetry of time, and the concept of the photon as an
entity which is not light yet plays a part in radiation
processes, to form a distinctive (but ignored) theory
of the nature of light.

522. Swenson, Loyd S., Jr. *The Ethereal Aether: A History of
          the Michelson-Morley-Miller Aether-Drift
          Experiments, 1880-1930.* Austin and London:
          University of Texas Press, 1972. xxii + 361
          pp. Appendices, Bibliography, Index.

   Describes in detail the experiments and the
experimenters. Argues that at the time it was contemplated
the experiment was not widely viewed as crucial, and that
it did not become truly significant until after the
acceptance of the Theory of Relativity made it very
useful as a means of explanation and confirmation of
Einstein's work.

523. Weart, Spencer R. "The Physics Business in America, 1919-
          1940: A Statistical Reconnaissance." *The
          Sciences in the American Context: New Perspec-
          tives* (item 84), pp. 295-358.

   Demonstrates that the physics community grew
very rapidly during this era. Finds that the Great
Depression had relatively little impact with the notable
exception of applied physics in industry and government.

524. Weiner, Charles Irwin. "Joseph Henry's Lectures on
          Natural Philosophy: Teaching and Research in
          Physics, 1832-1847." Ph.D. dissertation, Case
          Institute of Technology, 1965.

   Analyzes the college lectures of the leading
experimental physicist in mid-nineteenth-century America
within the contexts of the developments in physics and
the nature of the American college. Argues that Henry's
lectures represent scientific thinking out loud, giving
insights into his views on major issues in physics.

525. ————. "A New Site for the Seminar: The Refugees and American Physics in the Thirties." *The Intellectual Migration: Europe and America, 1930-1960.* Edited by Donald Fleming and Bernard Bailyn. Cambridge, Mass.: Harvard University Press, 1969, pp. 190-234.

Argues that the internationalism and mobility of the physics community in the 1920s and the self-improvement of the American physics community just prior to the immigration paved the way for the rapid and successful assimilation of the refugee physicists. Suggests that the social environment in America in the 1930s might have induced a "brain drain" of European physicists even without the stimulus of Hitler.

526. ————. "Physics in the Great Depression." *Physics Today*, 23 (October 1970): 31-37.

Finds that issues such as the social relevance and human implications of physics research were raised during this period, but not settled. Contends that the greatest impact of the Depression was upon young Ph.D.s looking for employment; established physicists were generally able to maintain employment although research funds had been cut sharply. Estimates that recovery had begun by 1935 and was complete by 1941.

527. ————. "Physics Today and the Spirit of the Forties." *Physics Today*, 26 (May 1973): 23-28.

Argues that when this journal emerged in 1948 it reflected the widespread optimism of the postwar physics community. Identifies this optimism with the excitement over discoveries in particle and solid state physics, the availability of new instruments and new sources of funding, and the public support resulting from the demonstrations of the applications of physics during World War II.

528. Wheeler, Lynde Phelps. *Josiah Willard Gibbs: The History of a Great Mind.* New Haven: Yale University Press, 1951. Reprint. Hamden: Archon Books, 1970. xiii + 270 pp. Appendices, Bibliography, Index.

Describes the career and accomplishments of what has been traditionally viewed as one of the major anomalies in the history of American science: a great nineteenth-century theorist in the physical sciences.

Discusses his fundamental contributions to physical
chemistry, thermodynamics, and statistical mechanics.

529. Williams, L. Pearce. "The Simultaneous Discovery of
               Electro-Magnetic Induction by Michael Faraday
               and Joseph Henry. *Bullétin de la société des
               amis d'André-Marie Ampère*, No. 22 (1965): 12-21.

        Shows that Faraday and Henry came to the same
discovery from diametrically opposed theoretical positions.
Henry depended upon the fluid theory of electricity.
Faraday, in contrast, had rejected electrical fluids
and was led to his successful experiments by his
suspicion that electricity was a wave phenomenon.

530. Winnik, Herbert C. "A Reconsideration of Henry A. Rowland
               --The Man." *Annals of Science*, 29 (1972):
               19-34.

        Contends that Rowland's scientific career was
greatly influenced by his personality and his response
to his image of his father as a dedicated minister.
Contradicts earlier biographical accounts which claim
little or no connection between Rowland's personal life
and his scientific endeavors.

CHAPTER IV: THE BIOLOGICAL SCIENCES

BIOLOGY AND THE BIOMEDICAL SCIENCES

531. Allen, Garland E.   "T.H. Morgan and the Emergence of a
             New American Biology." *The Quarterly Review
             of Biology*, 44 (1969): 168-188.

        Argues that Morgan symbolized the shift in
American biology at the end of the nineteenth century
from the descriptive and taxonomic to the experimental.
Traces Morgan's interest in experimentation in part to
his experiences at the Naples Marine Laboratory.

532. Atwater, Edward C.   "'Squeezing Mother Nature': Experi-
             mental Physiology in the United States Before
             1870." *Bulletin of the History of Medicine*,
             52 (1978): 313-335.

        Contends that it was the want of financial
support and the rudimentary state of the available technology, not a lack of interest, which retarded experimental
physiology in this country. Presents evidence of that
interest in the form of research conducted by medical
students for their dissertations.

533. Baxter, Alice Levine.   "E.B. Wilson's 'Destruction' of
             the Germ-Layer Theory." *Isis*, 68 (1977):
             363-374.

        Argues that Wilson was attacking the assumption
that homologous organs developed from the same germ-layer.
Concludes that he was unable to destroy the theory because
its complex nature made direct disproof very difficult.

534. ————.   "Edmund B. Wilson as a Preformationist: Some
                 Reasons for His Acceptance of the Chromosome
                 Theory." *Journal of the History of Biology*,
                 9 (1976): 29-57.

          Explores the preformation-epigenesis debate
     during the late nineteenth and early twentieth centuries.
     Concludes that Wilson's preformation position made it
     easier for him to accept the chromosome theory of inher-
     itance. Both theories shared the view that the egg was
     not homogeneous, but contained specific formative material;
     both theories also appealed to Wilson's morphological
     and materialist preferences.

535. Bylebyl, Jerome J. "William Beaumont, Robley Dunglison,
                 and the 'Philadelphia Physiologists.'" *Journal
                 of the History of Medicine*, 25 (1970): 3-21.

          Supplies the background for Beaumont's work,
     concentrating on the rejection of the chemical view of
     digestion by the Philadelphia community of physiologists.
     Argues that the resulting controversy led Beaumont to
     focus on the issues of the existence of gastric juice
     and the ability of that juice to dissolve food both
     within and outside the stomach.

536. Colin, Patrick L. "A Brief History of the Tortugas
                 Marine Laboratory and the Department of Marine
                 Biology, Carnegie Institution of Washington."
                 Oceanography: The Past. Edited by M. Sears
                 and D. Merriman. Proceedings of the Third
                 International Congress on the History of
                 Oceanography. New York, Heidelberg, and
                 Berlin: Springer-Verlag, 1980, pp. 138-147.

          Describes the activities of one of the major
     centers of tropical marine research prior to World War
     II. Credits the laboratory's ultimate decline to shifting
     interest towards cellular and molecular biology.

537. Corner, George W. *A History of the Rockefeller Institute,
                 1901-1953: Origins and Growth*. New York: The
                 Rockefeller Institute Press, 1964. xiii +
                 635 pp. Appendices, Index.

          Examines the development and contributions of
     one of the most important independent research institu-
     tions in America. Focuses on the tenure of its first
     two directors--Simon Flexner (1902-1935) and H.S. Gasser
     (1935-1953). Divides the narrative into three sections:

the founding and development through World War I; events
during the remainder of Flexner's tenure; the Rockefeller
under Gasser. For users of this bibliography, the detailed discussions of the Rockefeller's research in
physiology, chemistry, and genetics are the most
important aspects of this book.

538. Cravens, Hamilton. "The Role of Universities in the Rise
          of Experimental Biology." *The Science Teacher*,
          44 (1977): 33-37.

Argues that the United States became a center
for experimental biology during the period 1890-1910
because the discipline became identified with the new
graduate universities.

539. Evans, Mary Alice, and Howard Ensign Evans. *William
          Morton Wheeler, Biologist*. Cambridge, Mass.:
          Harvard University Press, 1970. xii + 363 pp.
          Bibliography, Index.

Interweaves the history of non-experimental
zoology in the United States during the late nineteenth
and early twentieth centuries with an account of Wheeler's
life. Presents Wheeler as a field naturalist with a
concern for the popular dissemination of scientific
knowledge, an interest in the philosophical issues in
science, but little sympathy for laboratory science,
especially genetics. Discusses the evolution of the
Bussey Institution of Harvard from a school of husbandry
and gardening to the Graduate School of Applied Biology,
a process in which Wheeler played a dominate role.

540. Fleming, Donald. "Émigré Physicists and the Biological
          Revolution." *The Intellectual Migration:
          Europe and America, 1930-1960*. Edited by
          Donald Fleming and Bernard Bailyn. Cambridge,
          Mass.: Harvard University Press, 1969,
          pp. 152-189.

Traces the intellectual history of the Crick-
Watson model of DNA, focusing on the contributions of
physicists to genetics and molecular biology. Finds
that these physicists appear to have been motivated by
their anxiety about the impact of the physical sciences
upon human life and their distaste for the "big science"
style which was coming to dominate physics.

541. ————. "Introduction." *The Mechanistic Conception of
            Life.* By Jacques Loeb. Cambridge, Mass.:
            Harvard University Press, 1964, pp. vii-xli.

    Describes the life and work of an individual who
blended the roles of philosopher, moralist, reformer, and
scientific investigator. Argues that he represented a
continuation of a well-established philosophical and
scientific tradition. Contends that the central issue
in his experiments was the existence of freedom of the
will.

542. Gussin, Arnold E.S. "Jacques Loeb: The Man and His
            Tropism Theory of Animal Conduct." *Journal of
            the History of Medicine and Allied Sciences,*
            18 (1963): 321-336.

    Argues that Loeb rejected criticism of his
theory of the physio-chemical mechanisms of behavior--
animals can be viewed as machines reacting to stimuli--as
based entirely on the teleological biases of his critics.

543. Hawkins, Hugh. "Transatlantic Discipleship: Two American
            Biologists and Their German Mentor." *Isis,*
            71 (1980): 197-210.

    Traces the growing independence of John M.
Tyler and Henry B. Ward from their mentor in experimental
zoology. Argues that World War I did not dissolve the
ties between the men.

544. Lillie, Frank R. *The Woods Hole Marine Biological
            Laboratory.* Chicago: University of Chicago
            Press, 1944. ix + 284 pp. Appendices, Index.

    Surveys the organizational, material, and
scientific development of this major center for basic
research in zoology, physiology, and embryology. Includes
much information, but little analysis.

545. Lurie, Edward. *Louis Agassiz: A Life in Science.* Chicago:
            University of Chicago Press, 1960. xiv + 449
            pp. Bibliography, Index.

    Describes the contributions in zoology, geology,
and paleontology of this Swiss-born naturalist who
immigrated to the United States in 1846. Discusses
Agassiz's role in the informal, elite network of American
scientists known as the Lazzaroni. Examines his efforts
to establish a great teaching museum, culminating in the
founding of the Museum of Comparative Zoology at Harvard.

Analyzes Agassiz's position as Darwin's chief American scientific opponent. Describes Agassiz's clashes with Asa Gray over evolution. This biography should be read in conjunction with item 556.

546. McCullough, Dennis M. "W.K. Brooks's Role in the History of American Biology." *Journal of the History of Biology*, 2 (1969): 411-438.

Rejects the traditional interpretation of Brooks as a great teacher, claiming that the quality of his students has inflated his reputation. Views Brooks as the last major figure in descriptive morphology at a time when the best American biologists were becoming experimentalists.

547. Maienschein, Jane Ann. "Ross Harrison's Crucial Experiment as a Foundation for Modern American Experimental Embryology." Ph.D. dissertation, Indiana University, 1980.

Analyzes Harrison's 1910 experiment on the process of nerve fiber development. Concludes that the experiment was not crucial because it provided definite proof or disproof of a particular theory but because it had fruitful consequences, serving as a model for a successful research program in experimental embryology.

548. Oppenheimer, Jane M. "Ross Harrison's Contributions to Experimental Embryology." *Bulletin of the History of Medicine*, 40 (1967): 525-543.

Observes that **his** specific contribution was the discovery of the technique of tissue culture; more generally, he demonstrated how to formulate specific questions which could be answered experimentally.

549. Parascandola, John. "Organismic and Holistic Concepts in the Thought of L.J. Henderson." *Journal of the History of Biology*, 4 (1971): 63-113.

Argues that the unifying themes in Henderson's research were his study of entire systems and his rejection of the thesis that the system was only the sum of its parts. Traces the development of these concepts and examines Henderson's application of them to biological problems.

550. Pauly, Philip Joseph. "Jacques Loeb and the Control of
              Life: An Experimental Biologist in Germany and
              America, 1859-1924." Ph.D. dissertation, Johns
              Hopkins University, 1980.

       Argues that Loeb's objective was the demonstra-
tion of science's power to control, manipulate, and
transform life; ultimately, he wanted to create life in
the laboratory. Claims that his mechanistic world view
was only a tool, not an end in itself.

551. Rosenberg, Charles E. "On the Study of American Biology
              and Medicine: Some Justifications." *Bulletin
              of the History of Medicine*, 38 (1964): 364-376.

       Advocates the study of the development of a
discipline within a national context as a means of over-
coming the internal-external historiographic split.
Argues that such studies offer the opportunity of relating
patterns of economic development to developments in
science, while also showing the social function of
scientific ideas. Contends that the study of a discipline
leads to modes of historical explanation superior to
those derived from the study of great men or schools of
scientific thought.

552. Sokatch, John R. "The Evolution of Molecular Biology."
              Issues and Ideas in America. Edited by Benjamin
              J. Taylor and Thurman J. White. Norman:
              University of Oklahoma Press, 1976, pp. 295-300.

       Cites improvements in instrumentation, increased
funding, and the immigration of European scientists as
major factors in the development of this discipline in
America.

553. Stephens, Lester D. "Joseph LeConte and the Development
              of the Physiology and Psychology of Vision in
              the United States." *Annals of Science*, 37
              (1980): 303-321.

       Evaluates LeConte's contributions to the study
of binocular vision. Finds that his book was the major
American text for over twenty years. Blames the limitations of his work on his unfamiliarity with some developments in German science and his neo-Lamarckian sympathies,
which resulted in some of his explanations placed in the
form of the accumulative effect of residual experiences.

554. Werdinger, Jeffrey. "Embryology at Woods Hole: The
                Emergence of a New American Biology." Ph.D.
                dissertation, Indiana University, 1980.

         Argues that the transformation of embryology
into an experimental science, primarily the result of
the activities of the scientists at the Marine Biological
Laboratory, required more than the introduction of new
methodologies; it also required a redefinition of the
goals of embryology. Attacks as oversimplifications the
theses that the new experimental biology resulted from
one particular line of development, or that there was a
clean break between the natural history traditions and
the experimentalists.

BOTANY

555. Berkeley, Edmund, and Dorothy Smith Berkeley. *John
                Clayton: Pioneer of American Botany.* Chapel
                Hill: The University of North Carolina Press,
                1963. x + 236 pp. Appendix, Bibliography,
                Index.

         Details the life and contributions of a
competent, thorough, but unspectacular member of the
Natural History Circle. Demonstrates how an international
correspondence network enabled scientists to collaborate
over long distances without ever having met. John Clayton
II (1694-1773, often confused with an earlier namesake)
published his catalogue of Virginia plants in the first
important British flora of North America.

*   Cittadino, Eugene. "Ecology and the Professionalization
                of Botany in America, 1890-1905." Cited above
                as item 216.

         Links the professionalization of American
botany to the development of the "new" botany, characterized by interest in all plants rather than just seed
plants and laboratory instruction. Discusses the work
of American botanists who investigated ecological problems
during this period, concluding that this research was
very uneven in quality and less sophisticated than that
in laboratory plant physiology.

556. Dupree, A. Hunter. *Asa Gray, 1810-1888.* Cambridge, Mass.:
                Harvard University Press, 1959. x + 505 pp.
                Index.

         Traces Gray's gradual transformation from
physician to professional botanist, focusing on the

informal educational opportunities afforded by the
American scientific community and the Grand Tour of
Europe. Describes Gray's role in making Harvard the
center of American botanical research and education, as
well as the clearinghouse for Western botanical collectors.
Analyzes Gray's contributions to Darwin's theory of
evolution through his observations of the similarities
between Asian and North American flora. Describes Gray's
defense of evolution, including his confrontations with
Louis Agassiz, and his efforts to be both a Darwinian
and a theist. This biography should be read in conjunction with item 545.

557. Ewan, Joseph. "Frederick Pursh, 1774-1820, and His
                Botanical Associates." *Proceedings of the
                American Philosophical Society*, 96 (1952):
                599-628.

        Views Pursh's *Flora* (1814) as the first inclusive
account of the plants of North America, marking the end
of the colonial era in American botany by bringing together
all the previous work and, at the same time, inspiring
a new generation of American botanists.

558. ———. "Plant Collectors in America: Backgrounds for
                Linneaeus." *Essays in Biohistory*. Edited by
                P. Smit and R.J. Ch. V. ter Laage. Utrecht:
                International Association for Plant Taxonomy,
                1970, pp. 19-54.

        Includes biographies of eighteenth-century
collectors and a very good bibliography.

559. ———, editor. *A Short History of Botany in the United
                States*. New York and London: Hafner Publishing
                Company, 1969. ix + 174 pp. Bibliography,
                Index.

        Contains a brief review of the institutional
context of botanical research in the United States and
brief surveys of the American contributions in thirteen
specialties or sub-disciplines, ranging from plant
physiology to medical botany. Most of the discussions
are limited to chronicles of significant events and names.

560. Humphrey, Harry Baker. *Makers of North American Botany*.
                New York: The Ronald Press Company, 1961.
                xi + 265 pp.

Presents brief outlines of the careers and contributions of 121 botanists. Relies primarily upon obituaries for information.

561. McVaugh, Rogers. "Botanical Results of the Michigan Geological Survey Under the Direction of Douglass Houghton, 1837-1840." *The Michigan Botanist*, 9 (1970): 213-243.

Finds that one result was the introduction of the previously unknown Michigan flora into the literature.

562. Overfield, Richard A. "Charles E. Bessey: The Impact of the 'New' Botany on American Agriculture, 1880-1910." *Technology and Culture*, 16 (1975): 162-181.

Presents Bessey as typical of his generation of agricultural scientists; his self-image was that of an experimental scientist not an agriculturist, yet his major contributions were in the application of science to agriculture, not basic research. Considers his most important role in American botany that of a clearinghouse for research.

563. Rodgers, Andrew Denny III. *American Botany, 1873-1892: Decades of Transition*. Princeton: Princeton University Press, 1944. Reprint. New York and London: Hafner Publishing Company, 1968. vii + 340 pp. Index.

Deals with the importation of European concepts and laboratory techniques and the resulting gradual evolution of botany from a descriptive to an experimental science. Discusses the rise of interest in paleobotany, spurred by the findings of the surveys of the West. Focuses on Asa Gray as the key figure in the transition, arguing that his death marked the beginning of the era of the domination of botanical research in America by the newer approaches.

564. ———. *John Merle Coulter: Missionary in Science*. Princeton: Princeton University Press, 1944. x + 321 pp. Index.

Presents not so much a biography as a history of an era constructed around the life of a single individual. Utilizes Coulter's career to examine the displacement of taxonomic descriptions by biochemical analyses in the conduct of botanical research. Offers Coulter as an

example of the scientist whose orthodox Christianity did
not prevent his acceptance of evolution. The same author
has written a number of other biographies which also ex-
plore larger issues in nineteenth and early twentieth-
century botany, including studies of William Starling
Sullivant (bryology), Liberty Hyde Bailey (horticulture),
and Bernhard Eduard Fernow (forestry).

565. ————. *John Torrey: A Story of North American Botany*.
        Princeton: Princeton University Press, 1942.
        Reprint. New York and London: Hafner Publishing
        Company, 1965. xii + 352 pp. Bibliography,
        Index.

Concentrates on the botanical activities and
accomplishments of this botanist, chemist, and mineralo-
gist. Describes his role in introducing the natural
system of classification into the United States. Delineates
his function as the clearinghouse for many of the botanical
specimens collected by frontier exploring and surveying
expeditions.

566. Stannard, Jerry. "Early American Botany and its Source."
        Bibliography of Natural History, 1966, pp.
        73-102.

Provides a comprehensive, although not complete,
survey of the primary and secondary source material available for the history of American botany prior to the
Civil War. Points out some valid needs.

567. Sutton, S.B. *Charles Sprague Sargent and the Arnold
        Arboretum*. Cambridge, Mass.: Harvard University
        Press, 1970. xvii + 382 pp. Index.

Explores the life of the first director of the
Arnold Arboretum (1873-1927). Demonstrates the importance
of personal wealth, influential friends, and social
position in successfully establishing a scientific
institution. Discusses the contributions of the
Arboretum to horticulture, especially through its
introduction of specimens from the Far East. Examines
Sargent's role in the national forestry and conservation
movements, tracing his gradual disenchantment with the
policies of the federal government.

EUGENICS AND GENETICS

568. Allen, Garland E. "Science and Society in the Eugenics
                Thought of H.J. Muller." *Bioscience*, 20
                (1970): 346-353.

        Examines Muller's work as an example of the
interaction of scientific ideas (gene action) and social
views (modification of human breeding). Argues that
Muller left the United States for the USSR in the 1930s
in hope of finding a supportive environment for his
program of evolving a noncompetitive society through
selective breeding.

569. ————. *Thomas Hunt Morgan: The Man and His Science*.
                Princeton: Princeton University Press, 1978.
                xvii + 447 pp. Appendices, Bibliography, Index.

        Utilizes Morgan's career as a vehicle for
exploring the history of genetics during the early
twentieth century and the transformation of American
biology from descriptive to experimental--the "revolt
from morphology." Describes the institutional settings
which played major roles in his career: the Marine
Biological Laboratory at Woods Hole; the "fly room" at
Columbia, where Morgan and his colleagues developed the
Mendelian-chromosome theory of heredity; and the Division
of Biological Sciences at the California Institute of
Technology. Discusses the important role of a supportive
family structure in the preservation of Morgan's scientific
productivity.

570. Brush, Stephen G. "Nettie M. Stevens and the Discovery
                of Sex Determination by Chromosomes." *Isis*,
                69 (1978): 163-172.

        Calls for a reexamination of the distribution
of credit between Stevens and E.B. Wilson. Asks historians
of biology to pay more attention to this relatively
neglected but significant figure.

571. Carlson, Elof Axel. "The *Drosophila* Group: The Transition
                from the Mendelian Unit to the Individual Gene."
                Journal of the History of Biology, 7 (1974):
                31-48.

        Describes the conflicting ideas and personali-
ties within the "fly group," the scientific unit which
provided major evidence for the theory of the gene.
Claims that the switch in genetics from the study of

inheritance and variation to the study of the inheritance
of variation was the result of the fly group's work.

572. Haller, John S., Jr. *Outcasts from Evolution: Scientific
         Attitudes of Racial Inferiority, 1859-1900.*
         Urbana, Chicago, and London: University of
         Illinois Press, 1971. xv + 228 pp. Bibliog-
         raphy, Index.

    Demonstrates that science provided a vocabulary
and a set of concepts which rationalized racial injustice
and verified racial inferiority and in doing so removed
the stigma of prejudice from efforts to segregate and
disenfranchise the Black population; science presented
such activities as the logical outgrowth of the demonstrated inferiority of Blacks.

573. Haller, Mark H. *Eugenics: Hereditarian Attitudes in
         American Thought.* New Brunswick: Rutgers
         University Press, 1963. vii + 264 pp.
         Bibliography, Index.

    Views eugenics as a scientific reform initiated
by elitist, socially conservative leaders to prevent the
domination of this country by what they perceived to be
social misfits. Divides the eugenics movement into
three stages: the belief that social misfits--including
paupers, criminals, the insane, and the feebleminded--
were the result of heredity took root during the years
1870-1905; from 1905 through 1930 eugenics was at the
height of its influence, but began to take racist over-
tones and evolve from a reform movement to a repudiation
and rejection of the American reforming tradition; post-
1930 eugenics lost support as new research illuminated
the complexity of human heredity, while the rise of
Hitler demonstrated the possible extremes of eugenic
activity.

574. Kevles, Daniel J. "Genetics in the United States and
         Great Britain, 1890-1930: A Review with
         Speculations." *Isis*, 71 (1980): 441-455.

    Argues that the present historiography has two
major defects: it neither discusses topics cross-
nationally nor explores the activities and attitudes
of non-elite members of the genetics community.

575. Ludmerer, Kenneth M. "American Geneticists and the
                Eugenics Movement, 1905-1935." *Journal of the
                History of Biology*, 2 (1969): 337-362.

        Links the declining support given social
legislation based on genetics by the American genetics
community to both internal developments in science which
increasingly demonstrated the complexities of heredity,
and external events, such as the exploitation of eugenics
by racists.

576. ————. *Genetics and American Society: A Historical
                Appraisal*. Baltimore and London: The Johns
                Hopkins University Press, 1972. xi + 222 pp.
                Bibliography, Index.

        Explores the utilization of genetic theories in
the determination of social policy and the impact of social
and political events on genetics between 1905 and 1930.
Focuses on the events leading to the Immigration Restric-
tion Act of 1924, when the eugenicists supplied the
scientific justification for the assessment of southern
and eastern Europeans as inferior, and hence undesirable,
immigrants. Concludes that scientific arguments will
not overcome popular prejudices but are effective tools
in social policy determination if science and popular
sentiment coincide.

577. Shine, Ian, and Sylvia Wrobel. *Thomas Hunt Morgan:
                Pioneer of Genetics*. Lexington: The University
                Press of Kentucky, 1976. xv + 160 pp. Index.

        Offers important insights into the human side
of Morgan: his relationship with his wife, a biologist
in her own right; his attitude towards students and
co-workers; and his indifference to political issues
even when they were of import to science. Skimps on the
descriptions and analysis of the content and context of
Morgan's contributions to science. Item 569 is superior
in this regard.

578. Vecoli, Rudolph. "Sterilization: A Progressive Measure?"
                Wisconsin Magazine of History, 43 (1960):
                190-202.

        Argues that the Progressives adopted a eugenic
solution for restricting the number of socially maladjusted
individuals because such an approach appealed to both the
basic tenets of Progressivism and the biases of the
generally middle-class Progressives. Describes

sterilization as a scientific, humane solution which
sacrificed the individual for the good of society, while
sanctioning the established social order.

NATURAL HISTORY

579. Alden, Roland H., and John D. Ifft, III. *Early Naturalists
        in the Far West.* Occasional Papers of the
        California Academy of Sciences, XX. San
        Francisco: California Academy of Sciences,
        1943. 59 pp. Bibliography.

        Includes discussions of the scientific contri-
butions made by European expeditions to the region.
Terminates at the point when scientific activity was
becoming indigenous.

*   Aldrich, Michele. "New York Natural History Survey,
        1836-1845." Cited above as item 278.

        Provides a detailed description of the survey's
operation and results.

580. Berkeley, Edmund, and Dorothy Smith Berkeley. *Dr.
        Alexander Garden of Charles Town.* Chapel
        Hill: University of North Carolina Press,
        1969. xv + 379 pp. Appendices, Bibliography,
        Index.

        Describes the life of the Charleston physician
who was one of the few colonial contributors to zoology.
Acclaims him as one of the leading systematists in the
colonies. Discusses the events leading to his expulsion
as a Loyalist during the Revolutionary War.

581. ———— and ————. *Dr. John Mitchell: The Man Who Made
        the Map of North America.* Chapel Hill:
        University of North Carolina Press, 1974.
        xix + 283 pp. Bibliography, Index.

        Assesses the scientific contributions of a
minor figure in natural history, although he was the
first North American to write on taxonomic principles.
Provides a detailed account of the publication of his
1755 map of North America, produced to point out the
threat French expansion presented to the British.

582. Burroughs, Raymond Darwin, editor. *The Natural History of the Lewis and Clark Expedition.* East Lansing: Michigan State University Press, 1961. xii + 340 pp. Appendix, Index.

Extracts the zoological observations from the expedition's journals. Demonstrates how slowly the American scientific community assimilated this information.

583. Cutright, Paul Russell. *Lewis and Clark: Pioneering Naturalists.* Urbana: University of Illinois Press, 1969. xvi + 506 pp. Appendices, Bibliography, Index.

Summarizes the contributions of the expedition to cartography, meteorology, ethnology, botany, and zoology. Contends that Lewis was one of the leading American naturalists of his day. Offers no explanation for the failure of the American scientific community to exploit more fully the findings of the two men. Represents a superior work to item 582.

584. Cutting, Rose Marie. *John and William Bartram, William Byrd II and St. John de Crevecoeur: A Reference Guide.* Boston: G.K. Hall and Company, 1976. xxiii + 174 pp. Index.

Presents a descriptive, non-evaluative, and very extensive guide to secondary sources.

585. Ewan, Joseph. *Rocky Mountain Naturalists.* Denver: The University of Denver Press, 1950. xiv + 358 pp. Bibliography, Index.

Includes nine detailed biographical studies and a roster of approximately eight hundred natural history collectors active in the Colorado region of the Rockies during the years 1682-1932.

586. ———, and Nesta Ewan. *John Banister and His Natural History of Virginia, 1678-1692.* Urbana: University of Illinois Press, 1970. xxx + 485 pp. Chronology, Bibliography, Index.

Examines the life and scientific accomplishments of one of the first significant resident American naturalists. Explores the European support system which was essential to his activities. Includes reproductions of Banister's catalogues, other natural history writings, and drawings.

587. Geiser, Samuel Wood. *Naturalists of the Frontier*. 2nd
         edition. Dallas: Southern Methodist University,
         1948. 296 pp. Appendix, Bibliography, Index.

         Examines natural history collecting in Texas
    between 1820 and 1880. Chooses this period and locale
    because of the coincidence of the scientific and social
    frontiers. Provides detailed discussions of eleven
    naturalists, almost all European born and educated, who
    came to Texas either specifically to collect specimens
    or for the economic opportunities and political freedom
    available on the frontier, and remained long enough to
    make important contributions. Includes short descriptions
    of over 150 additional naturalists and collectors active
    in the same region during this period.

588. Graustein, Jeanette E. *Thomas Nuttall, Naturalist:
         Explorations in America, 1808-1841*. Cambridge,
         Mass.: Harvard University Press, 1967. xiv +
         481 pp. Index.

         Documents the life and contributions of an
    important field naturalist. Presents a sympathetic but
    scholarly evaluation of Nuttall's work in both botany and
    zoology, his pedagogical efforts at Harvard, and his
    explorations. Fails, however, to catch all the nuances
    of the developing conflict between field naturalists
    and the closet botanists like Asa Gray.

589. Hellman, Geoffrey. *Bankers, Bones & Beetles: The First
         Century of The American Museum of Natural
         History*. Garden City, N.Y.: The Natural
         History Press, 1969. 275 pp. Index.

         Offers an informal history, devoid of scholarly
    apparatus. Concentrates on the personalities of the
    staff, especially the more colorful ones. Shows little
    interest in the larger cultural and scientific contexts
    within which the museum was embedded.

590. Kastner, Joseph. *A Species of Eternity*. New York:
         Alfred A. Knopf, 1977. xii + 350 pp.
         Bibliography, Index.

         Surveys natural history in America from the
    mid-eighteenth to the mid-nineteenth centuries. Empha-
    sizes the adventure of field collecting and the interaction
    of diverse personalities. Traces the eventual eclipse
    of the field naturalist by the closet scientist.

591. Kennedy, John M. "Philanthropy and Science in New York
City: The American Museum of Natural History,
1868-1968." Ph.D. dissertation, Yale University, 1968.

    Argues that the businessmen who provided the
financial support for the museum were motivated by
desires to increase the city's prestige and provide
opportunity for public education. Investigates the
relationship between the values of the business community
and the content of scientific investigations conducted
at the museum; finds only superficial influences.

592. Kerkkonen, Martti. *Peter Kalm's North American Journey:
Its Ideological Background and Results*.
Helsinki: Finnish Historical Society, 1959.
260 pp. Bibliography, Index.

    Examines the eighteenth-century Swedish ideology
that everything in nature was for the amusement, profit,
or service of man. Finds that the purpose of Kalm's
explorations was the discovery and importation of
economically important plants; i.e., useful in manufacturing, food supply, or medication.

593. Poesch, Jessie. *Titian Ramsey Peale 1799-1885 and His
Journals of the Wilkes Expedition*. Memoirs of
the American Philosophical Society, 52.
Philadelphia: American Philosophical Society,
1961. x + 214 pp. Bibliography, Index.

    Concentrates on Peale's career as a naturalist-
artist and the supercedure of the naturalist-field
collector by the closet scientist who excelled in
technical classification and description. Deals only
superficially with Peale's life after the suppression
of his zoological volume by Wilkes.

594. Porter, Charlotte M. "The Concussion of Revolution:
Publications and Reform at the Early Academy
of Natural Sciences, Philadelphia, 1812-1842."
Journal of the History of Biology, 12 (1979):
273-292.

    Studies a group of nationalistic field natural-
ists, artists, engravers, and printers active during the
early years of the Academy whose cooperative efforts
resulted in the production of expensive, illustrated
natural history books aimed at general audiences. Argues
that their ignorance of European developments eventually
led to their downgrading by other American scientists.

595. ———. "'Subsilentio': Discouraged Works of Early
         Nineteenth-Century American Natural History."
         Journal of the Society for the Bibliography of
         Natural History, 9 (1979): 109-119.

     Contends that the work of certain field natural-
ists was suppressed because they were "splitters,"
distinctly nationalistic, and lacked knowledge of the
scientific literature at a time when the opposite
characteristics were in favor. Views this suppression
as part of the trend, completed by 1835, which reduced
the status of field naturalists from full-fledged members
of the scientific community to auxiliaries operating for
the benefit of resident or "closet" naturalists.

596. Sellers, Charles Coleman. *Mr. Peale's Museum: Charles*
         Willson Peale and the First Popular Museum of
         Natural Science and Art. New York: W.W. Norton
         & Company, 1980. xiv + 370 pp. Index.

     Shows that the museum fulfilled the three major
functions of a modern museum: the exhibition and
preservation of specimens, the support of research,
and the diffusion of knowledge through entertainment.
Traces the decline of the museum from a center for the
advancement of science to an entertainment business.

597. Smallwood, William M., and Mabel S.C. Smallwood. *Natural*
         History and the American Mind. New York:
         Columbia University Press, 1941. Reprint.
         New York: AMS Press, 1967. xiii + 445 pp.
         Bibliography, Index.

     Represents an extremely early attempt at
describing and analyzing the relationship between
American civilization and natural history. Limits
itself to the eastern part of the nation prior to the
rise of modern evolutionary science and the replacement
of the naturalists by specialized scientists, circa
1850. Stands as an important historiographic document.

598. Sterling, Keir Brooks. *Last of the Naturalists: The*
         Career of C. Hart Merriam. 2nd edition.
         New York: Arno Press, 1977. xii + 472 pp.
         Bibliography, Index.

     Traces the career of the first head of what is
now the United States Fish and Wildlife Service. Focuses
on his government service and his contribution to
mammalogy. Concludes that his strength as a scientist

was the thoroughness of his field research; his weakness was his refusal to modify positions or cope with objections. Finds that he was a poor administrator and lobbyist, clashing frequently with Congress over appropriations. Contains a detailed bibliography of great value.

599. Warner, Deborah Jean. *Graceanna Lewis: Scientist and Humanitarian*. Washington, D.C.: Smithsonian Institution Press, 1979. 139 pp. Index.

Studies the life of a representative woman amateur naturalist of the last half of the nineteenth century. Documents the difficulties she faced in obtaining acceptance by the professional scientific community but fails to explain why she was ultimately unable to establish a scientific career--was it her refusal to accept the standards of behavior of the predominately male community towards her sex, or the philosophical orientation of her scientific efforts?

600. Wilson, David Scofield. *In the Presence of Nature*. Amherst: University of Massachusetts Press, 1978. xix + 234 pp. Bibliography, Index.

Analyzes the role of nature reporters (observers and collectors of natural history specimens) in legitimatizing and domesticating nature in colonial America. Argues that nature was problematic to colonial settlers; the nature reporters investigated the wilderness and explained it in terms of the ongoing culture. Characteristics of nature reporters include a broad appreciation of nature, attention to detail, the utilization of vivid descriptions, selective use of the vernacular, and the ability to infer the message of nature. Provides three examples: Jonathan Carver, the traveller as nature reporter; John Bartram, the scientist as nature reporter; and Mark Catesby, the artist-naturalist as nature reporter. Compares all three with their European counterparts and demonstrates how the colonials were involved with their subject, while the Europeans simply observed it.

PALEONTOLOGY

601. Gerstner, Patsy A. "Vertebrate Paleontology, an Early Nineteenth-Century Transatlantic Science." *Journal of the History of Biology*, 3 (1970): 137-148.

Argues that Europeans began to respect and
utilize American descriptive and interpretive studies
after Americans adopted Cuvier's theoretical framework
circa 1830. Finds that this reversed the earlier attitude
that Americans were only collectors.

602. Gregory, Joseph T. "North American Vertebrate Paleontology,
       1776-1976." *Two Hundred Years of Geology in
       America: Proceedings of the New Hampshire
       Bicentennial Conference on the History of
       Geology* (item 461), 305-335.

Argues that there was little activity prior to
1850, tremendous individual effort from 1850 through 1890,
and a concern for the establishment and development of
museums as centers of research post-1890. Concentrates
on the contributions of individuals usually ignored by
historians preoccupied with the struggle between Cope
and Marsh; e.g., John Strong Newberry.

603. Howard, Robert West. *The Dawnseekers: The First History
       of American Paleontology.* New York and London:
       Harcourt Brace Jovanovich, 1975. xiii +
       314 pp. Bibliography, Index.

Limits its examination to the development of
vertebrate paleontology. Places the story within a
historically untenable framework of a continuing conflict
between science and theology. This work is factually
unreliable and undocumented.

604. Lanham, Uri. *The Bone Hunters.* New York: Columbia
       University Press, 1973. xi + 285 pp.
       Bibliography, Index.

Presents a semi-popular account of vertebrate
paleontology in the United States. Concentrates on the
activities of Cope, Marsh, and Joseph Leidy. Offers
little new information but is a reasonably accurate and
well-written account.

605. Osborn, Henry Fairfield. *Cope: Master Naturalist.*
       Princeton: Princeton University Press, 1931.
       Reprint. New York: Arno Press, 1978.
       xvi + 740 pp. Bibliography.

Offers a detailed and sympathetic rendering of
the life and work of Edward Drinker Cope, America's
leading naturalist during the last third of the nineteenth
century. Analyzes Cope's role in the rise of neo-Lamarck-

ianism in the United States. Must be read in conjuction
with item 606, the biography of Cope's arch-rival in
vertebrate paleontology.

606. Schuchert, Charles, and Clara Mae Levene. *O.C. Marsh:*
Pioneer in Paleontology. New Haven: Yale
University Press, 1940. Reprint. New York:
Arno Press, 1978. Appendix, Bibliography, Index.

Offers a detailed and sympathetic account of
the personal and scientific sides of Marsh; weak on
Marsh's relationship with the United States Geological
Survey. This biography was written in response to item
605. It remains, like item 605, a fundamental starting
point for understanding vertebrate paleontology in the
United States.

607. Shor, Elizabeth Noble. *Fossils and Flies: The Life of a*
Compleat Scientist, Samuel Wendell Williston
(1851-1918). Norman: University of Oklahoma
Press, 1971. xiv + 285 pp. List of Publications,
Bibliography, Index.

Relates the life of a paleontologist and
entomologist who began his career as a collector for
Marsh. Throws light on the hostile relationship between
Marsh and his junior staff due to Marsh's restrictions
on their independent research.

608. ————. *The Fossil Feud: Between E.D. Cope and O.C. Marsh.*
Hicksville, New York: Exposition Press, 1974.
xi + 340 pp. Bibliography, Index.

Focuses on the public airing of the personal
and professional conflict between Cope and Marsh in the
New York *Herald* during January 1890. Reprints the newspaper columns, explains the background, and offers a
detailed analysis. Includes short biographical sketches
and precis of the involvement of the nearly one hundred
individuals named in the newspaper accounts.

609. Simpson, George Gaylord. "The Beginnings of Vertebrate
Paleontology in North America." *Proceedings of*
the American Philosophical Society, 86 (1942):
130-188.

Surveys activity in American vertebrate paleontology until the end of the pioneer period, set arbitrarily
at 1842. Emphasizes the central role of the American
Philosophical Society in the growth of the discipline
through its meetings, publications, and collections.

ZOOLOGICAL DISCIPLINES

610. Ainley, Marianne Gosztonyi. "The Contribution of the
            Amateur to North American Ornithology: A
            Historical Perspective." *The Living Bird*,
            18 (1979-80): 161-177.

        Demonstrates the slow pace of professionaliza-
    tion of ornithology; professional domination of the
    discipline did not occur until the second third of the
    twentieth century.

611. Allen, Elsa G. "The History of American Ornithology
            Before Audubon." *Transactions of the American
            Philosophical Society*, n.s., 41 (1951): 385-
            591. Reprint. New York: Russell and Russell,
            1969.

        Discusses traveller's accounts, bird lists,
    anatomical studies, and the occasional naturalist's
    descriptions. Provides an excellent narrative account
    although lacks analysis at certain points; e.g., the
    cessation around 1730 and subsequent resumption after
    1750 of colonial studies of birds is described but not
    explained.

612. Benson, Norman G., editor. *A Century of Fisheries in
            North America*. Special Publication No. 7.
            Washington, D.C.: American Fisheries Society,
            1970. x + 298 pp.

        Consists of relatively short articles focusing
    on research either in particular specializations (e.g.,
    the cultivation of particular classes of fish) or en-
    vironments (e.g. ponds). Contains item 618.

613. Cantwell, Robert. *Alexander Wilson: Naturalist and
            Pioneer*. Philadelphia and New York: J.B.
            Lippincott Company, 1961. 319 pp. Appendices,
            Bibliography, Index.

        Describes the life of the Scottish poet and
    satirist who, during less than a decade of intense
    effort, revolutionized American ornithology.

614. Ford, Alice. *John James Audubon*. Norman: University of
            Oklahoma Press, 1964. xiv + 488 pp.
            Chronology, Bibliography, Index.

        Offers a very comprehensive, detailed, balanced
    account of the great naturalist-artist, successor to

Alexander Wilson in the creation of accurate ornithological
illustrations. Provides new information about Audubon's
ancestry and youth.

615. Frick, George Frederick, and Raymond Phineas Stearns.
               Mark Catesby: The Colonial Audubon. Urbana:
               University of Illinois Press, 1961. x +
               137 pp. Bibliography, Appendix, Index.

        Argues that Catesby was the most significant
collector sent over to the New World by England during
the colonial era. Describes his book as the most outstanding natural history of a British possession in North
America published pre-1776. Contends that Catesby's
greatest impact was in ornithology, where he provided the
first reasonably accurate published pictorial accounts.
Finds that his classifications were made obsolete quite
rapidly by the work of Linnaeus.

616. Hubbs, Carl L. History of Ichthyology in the United
               States After 1850." *Copeia*, 1964, pp. 42-60.

        Summarizes all the research conducted between
1850 and 1925 upon fish native to North American waters
and by Americans throughout the world. Observes that by
1900 most descriptive work on American fish was thought
complete; American ichthyologists then turned to the
waters of the Pacific and off Latin America for new
specimens.

617. Hume, Edgar Erskine. *Ornithologists of the United States
               Army Medical Corps: Thirty-six Biographies*.
               Baltimore: The Johns Hopkins Press, 1942.
               Reprint. New York: Arno Press, 1978. xxv +
               583 pp. Index.

        Offers highlights of the careers and contribu-
tions of men who utilized their time in isolated and
unexplored locations to further science. Depends in
part upon archival research. Includes lists of birds
named by each ornithologist.

618. McHugh, J.L. "Trends in Fishery Research." *A Century
               of Fisheries in North America* (item 612),
               pp. 25-56.

        Contends that freshwater and marine fishery
research have developed at different rates and in slightly
different directions but that the overall trends have
been similar. Emphasizes the history of marine fishery

research. Argues that the policy of the United States
Fish Commission to accentuate fishery culture at the
expense of basic knowledge has inhibited the growth of
scientific knowledge of marine fishery.

619. Mallis, Arnold. *American Entomologists*. New Brunswick:
         Rutgers University Press, 1971. xvii + 549 pp.
         Index.

         Offers individual biographical sketches based
almost exclusively upon obituaries. Discusses careers
and lives but presents relatively little information about
scientific accomplishments. Divides the scientists along
disciplinary lines but includes separate groupings for
educators and state and federal entomologists.

620. Mayr, Ernst. "Materials for a History of American
         Ornithology." *Ornithology from Aristotle to
         the Present*. By Erwin Stresemann. Cambridge,
         Mass. and London: Harvard University Press,
         1975, pp. 365-396, 414-419.

         Calls for a history of the last century of
American ornithology. Surveys the contributions of
various institutions and summarizes the accomplishments
of both professional and amateur ornithologists.

621. Myers, George S. "A Brief Sketch of the History of
         Ichthyology in America to the Year 1850."
         Copeia, 1964, pp. 33-41.

         Demonstrates that American fish fauna became
a subject of intense regional interest after the Revolutionary War. Observes that the coastal waters of New
England and New York were one of only three areas in
the world circa 1850 where most of the fish fauna had
been discovered and described.

622. Welker, Robert Henry. *Birds and Men: American Birds in
         Science, Art, Literature, and Conservation,
         1800-1900*. Cambridge, Mass.: Harvard University
         Press, 1955. 230 pp. Bibliography, Index.

         Warns of the widespread sacrifice of accuracy
for style and anthromorphism in most American bird art
and literature. Identifies and evaluates Audubon's and
Thoreau's contributions as the first Americans to synthesize art and accuracy in painting and prose respectively.
Proves especially valuable for its discussions of the use
of birds in American art, prose, poetry, and fashion; less
valuable in delineating the development of ornithology.

CHAPTER V: THE SOCIAL SCIENCES

GENERAL

623. Anderson, C. Arnold. "The Striving for Cooperative
            Autonomy: American Social Sciences Over Fifty
            Years." *Social Forces*, 29 (1950): 8-19.

    Finds that the boundaries between the social
sciences (anthropology, economics, sociology, and political
science) had become more fixed between 1900 and 1949
although the transfer of ideas from one social science
to another increased during the same period.

624. Cravens, Hamilton. *The Triumph of Evolution: American
            Scientists and the Heredity-Environment Con-
            troversy, 1900-1941.* Philadelphia: University
            of Pennsylvania Press, 1978. xvi + 351 pp.
            Bibliography, Index.

    Finds that during the years 1890-1920 experimental
biologists and psychologists generally believed that
nature (heredity) was more important than nurture
(environment or culture) in molding human development,
in contrast to the anthropologists and sociologists, who
emphasized the role of nurture. Studies debates within
the scientific community over eugenics, the role of
human instincts, and mental testing. Concludes that by
World War II the American scientific community had accepted
the concept of the interdependency of heredity and
environment--inheritance provided the potential, while
the environment governed the actuality.

625. Davis, Robert C. "The Beginnings of American Social
            Research." *Nineteenth Century American Science:
            A Reappraisal* (item 32), pp. 152-178.

    Studies the antebellum efforts at collecting
social statistics for the purposes of providing necessary
data for intelligent legislation, accurate insurance

rates, and social reform activity. Argues that this
pioneering social research stressed the utility of data,
while staying away from grand theorizing or utopian
schemes.

626. Deegan, Mary Jo, and John S. Burger. "George Herbert
            Mead and Social Reform: His Work and Writings."
            *Journal of the History of the Behavioral
            Sciences*, 14 (1978): 362-373.

Demonstrates the invalidity of two theses--that
Mead published little during his lifetime and that *Mind,
Self and Society* was his most important work--by studying
his many attempts to use science for the betterment of
society.

627. Furner, Mary O. *Advocacy and Objectivity: A Crisis in
            the Professionalization of American Social
            Science, 1865-1905*. Lexington: The University
            Press of Kentucky, 1975. xv + 356 pp.
            Bibliography, Index.

Examines the processes of the differentiation
and professionalization of the social scientist in light
of the two conflicting roles available for these scientists
--social advocacy and academic objectivity. Focuses on
the economists as they develop standards of professional
behavior which forbade political or moral advocacy.
Analyzes the refinement which those standards underwent
during the academic freedom cases at the end of the
century. Includes political science and sociology within
the discussion but excludes history and anthropology.

628. Haskell, Thomas L. *The Emergence of Professional Social
            Science: The American Social Science Association
            and the Nineteenth-Century Crisis of Authority*.
            Urbana, Chicago, and London: University of
            Illinois Press, 1977. xii + 276 pp.
            Bibliography, Index.

Argues that the American Social Science
Association (1865-1909) was part of the movement by the
gentry to re-legitimize their authority by reestablishing
it upon a new basis--competence in social science. Finds
that the members of the ASSA--a coalition of natural
scientists, professionals, and education reformers--
believed that the social problems created by industrialization could be solved if there was only enough data.
Claims that the ASSA died because the specialized

professional societies created after 1883 absorbed its membership. Views the specialized societies as repre- senting the younger generation of professional scientists who operated under a completely different set of assumptions regarding the causality of social problems and their solution than the founders of the ASSA.

629. Lyons, Gene M. *The Uneasy Partnership: Social Science and the Federal Government in the Twentieth Century.* New York: Russell Sage Foundation, 1969. xvi + 394 pp. Appendices, Index.

Credits the increasing role of the social sciences in the federal government to the growing influence of the government in social and economic life, internal developments of administrative techniques for the appli- cation of the social sciences. Identifies three unsettled questions regarding this relationship: Can the social sciences contribute effectively to the formation of public policy? Should the social sciences seek federal support? Does the involvement in government, including the acceptance of financial support, mean the loss of scientific independence?

630. Mueller, Ronald H. "A Chapter in the History of the Relationship Between Psychology and Sociology in America: James Mark Baldwin." *Journal of the History of the Behavioral Sciences*, 12 (1976): 240-253.

Reviews the theoretical and organizational contributions of Baldwin to the integration of biological, sociological, and psychological principles into a new approach to the study of humans and their interaction with their society (social psychology).

631. Ross, Dorothy. "The Development of the Social Sciences." *The Organization of Knowledge in Modern America, 1860-1920*. Edited by Alexandra Oleson and John Voss. Baltimore and London: The Johns Hopkins University Press, 1979, pp. 107-138.

Focuses on the factors which fueled the evolu- tion of the fields of psychology, anthropology, economics, sociology, and political science into separate, academic, professional, and scientific disciplines.

ANTHROPOLOGY

632. Bieder, Robert E.   "Albert Gallatin and the Survival of
            Enlightenment Thought in Nineteenth-Century
            American Anthropology." *Towards a Science of
            Man: Essays in the History of Anthropology*.
            Edited by Timothy H.H. Thoresen. The Hague
            and Paris: Mouton Publishers, 1975, pp. 91-98.

        Observes that Gallatin always upheld the
Enlightenment tenet that man progresses through time,
contradicting the generally accepted attitude of contemporary mid-nineteenth-century anthropologists that the
American Indian had not evolved.

633. ————.   "The American Indian and the Development of
            Anthropological Thought in the United States,
            1780-1851." Ph.D. dissertation, University
            of Minnesota, 1972.

        Studies the work of five ethnologists--Benjamin
Smith Barton, Albert Gallatin, Henry Rowe Schoolcraft,
Ephraim George Squier, and Lewis Henry Morgan--from the
perspectives of their respective cultural values, the
alternatives available, and the data and sources underlying their work.

634. Darnell, Regna.   "The Professionalization of American
            Anthropology: A Case Study in the Sociology of
            Knowledge." *Social Science Information*, 10,
            No. 2 (April 1971): 83-103.

        Argues that the social organization of science
affects the scientific research. Finds two distinct
paradigms in the professionalization and organization of
American anthropology: the Bureau of American Ethnology,
led by John Wesley Powell, which stressed practical
justification and integrative, cooperative research;
and Franz Boas's concept of academic anthropology, which
called for basic, individualistic research.

635. Ewers, John C.   "William Clark's Indian Museum in St.
            Louis, 1816-1838." *A Cabinet of Curiosities:
            Five Episodes in the Evolution of American
            Museums* (item 103), pp. 49-72.

        Describes a proprietory museum which did not
survive the death of its owner.

636. Gruber, Jacob W. "Horatio Hale and the Development of
             American Anthropology." *Proceedings of the
             American Philosophical Society*, 111 (1967):
             5-37.

         Rejects the thesis that Boas created anthropology
in the United States. Claims Hale initiated Boas into the
traditions of the already well-established community of
American practitioners. Identifies those characteristics
of Boas's work which were derived from Hale.

637. Hart, Kevin Robert. "Government Geologists and the Early
             Man Controversy: The Problem of 'Official'
             Science in America, 1879-1907." Ph.D.
             dissertation, Kansas State University, 1976.

         Examines the late-nineteenth-century hypothesis
of the existence of pre-Indian cultures in America. Finds
that the leading critics of the idea were government
scientists (employed at the Geological Survey and Bureau
of American Ethnology), elevating their criticism to a
government position. Discusses the hostility and jealousy
proponents of the hypothesis felt towards these government
scientists, who had the power, the resources, and the
vehicles to propagate their point of view.

638. Hinsley, Curtis M., Jr. "The Development of a Profession:
             Anthropology in Washington, D.C., 1846-1903."
             Ph.D. dissertation, University of Wisconsin-
             Madison, 1976.

         Studies the development of anthropology in
Washington, D.C. from the establishment of the Smithsonian
to the death of John Wesley Powell. Focuses on the
activities of Joseph Henry and Otis T. Mason at the
Smithsonian, and Powell and W.J. McGee at the Geological
Survey and Bureau of American Ethnology. Argues that the
professionalization of anthropology was a result of
these activities.

639. ————, and Bill Holm. "A Cannibal in the National Museum:
             The Early Career of Franz Boas in America."
             American Anthropologist, 78 (1976): 306-316.

         Finds that Boas used cooperative arrangements
between institutions to compensate for his lack of a
strong institutional base.

640. Horsman, Reginald. "Scientific Racism and the American
         Indian in the Mid-Nineteenth Century."
         American Quarterly, 27 (1975): 152-168.

         Demonstrates how the arguments of the American
     School of Anthropology regarding the inherent inferiority
     of the American Indian became the accepted wisdom in the
     1840s and a justification for Indian policy and American
     expansion.

641. Judd, Neil M. *The Bureau of American Ethnology: A*
         Partial History. Norman: University of Oklahoma
         Press, 1967. xi + 139 pp. Index.

         Offers a summary analysis of the leadership,
     staff, collaborators, and publications of the BAE. Argues
     that the early clashes between the Smithsonian and the
     BAE represented policy disputes over priorities in the
     distribution of funds for ethnology--museum collections
     or field research. Concludes that the Smithsonian viewed
     the BAE as an "unwanted but tolerated stepchild."

642. Karcher, Carolyn L. "Melville's 'The 'Gees': A Forgotten
         Satire on Scientific Racism." *American Quarterly*,
         27 (1975): 421-442.

         Assesses this very short story as a parody of
     the scientific treatises on Blacks written by members of
     the American School of Anthropology.

643. Lurie, Edward. "Louis Agassiz and the Races of Man."
         Isis, 45 (1954): 227-242.

         Contends that Agassiz provided the theory of
     racial inferiority developed by the American School of
     Anthropology with scientific stature. Suggests that
     Agassiz switched from a monogenetic to a polygenetic
     position in part because of his general belief in the
     independent creation of species, and in part as a
     response to his first contact with Blacks and his
     perception of distinct racial differences.

644. Mark, Joan. "Frank Hamilton Cushing and an American
         Science of Anthropology." *Perspectives in*
         American History, 10 (1976): 447-486.

         Rejects the thesis that American anthropology
     borrowed its formative methods and theories from Europe.
     Examines the career of one counter-example: the developer
     of the modern concept of culture, with its assumptions
     of relativism, plurality, and wholeness.

645. Modell, Judith. "Ruth Benedict, Anthropologist: The
Reconciliation of Science and Humanism."
*Toward a Science of Man: Essays in the History
of Anthropology.* Edited by Timothy H.H. Thoresen.
The Hague and Paris: Mouton Publishers, 1975,
pp. 183-203.

Argues that anthropology provided a psycholog-
ically satisfying cultural role for women between the
wars. The science combined the insights of literature
with the certainties of science, while offering a career
--Benedict's being an examplar--in which women could
reach their potential and find rewarding experiences
without violating cultural expectations or accepted
behavior patterns.

646. Popkin, Richard H. "Pre-Adamism in 19th Century American
Thought: 'Speculative Biology' and Racism."
Philosophia, 8 (1978): 205-239.

Traces the evolution of the theory of the
existence of human life before Adam. Concludes that the
version developed by the American School of Anthropology
was the most powerful scientifically. Argues that the
racist implications of pre-Adamite theories were also
brought to their highest fruition in America.

647. Ray, Verne F., and Nancy Oestreich Lurie. "The Contributions of Lewis and Clark to Ethnography."
Journal of the Washington Academy of Sciences,
44 (1954): 358-370.

Argues that their work was generally ignored by
later generations of anthropologists because of a lack of
interest in the ethnohistorical methods necessary to
exploit their findings and the confusion and ambiguity
in their identification of tribes.

648. Resek, Carl. *Lewis Henry Morgan: American Scholar.*
Chicago: University of Chicago Press, 1960.
xi + 184 pp. Bibliography, Index.

Traces the waxing and waning influence of
Morgan's thought. Argues that he was at intellectual
peace with his own society; his ideas subsequently fell
into disfavor as changing social conditions and new
philosophical currents raised questions about his
assumptions and laws. Describes Morgan's attempts to
preserve Indian life. Credits his *League of the Iroquois*
as the first description of Indians to be generally free

of value judgements. Contends that Boas's approach illuminated the extent of Morgan's ethnocentrism.

649. Rogge, A.E. "A Look at Academic Anthropology: Through a Graph Darkly." *American Anthropologist*, 78 (1976): 829-843.

Tests the applicability of the theories utilized to analyze the history of the physical sciences (e.g., those of Kuhn and Price) to anthropology.

650. Spencer, Frank. "Two Unpublished Essays on the Anthropology of North America by Benjamin Smith Barton." *Isis*, 68 (1977): 567-573.

Spotlights the nationalistic and polygenistic elements in separate essays on the North American Indian and the albino.

651. Speth, William W. "The Anthropogeographic Theory of Franz Boas." *Anthropos*, 73 (1978): 1-31.

Investigates Boas's views on the influence of the environment upon culture. Finds that he rejected geographic determinism in favor of a complex, synthetic theory which considered the environment a limiting and modifying factor in the development of culture.

652. Stanton, William. *The Leopard's Spots: Scientific Attitudes Towards Race in America, 1815-59.* Chicago and London: University of Chicago Press, 1960. ix + 245 pp. Index.

Traces the rise and fall of the American School of Anthropology, which had theorized that various human races were distinct species created separately. Concludes that these scientists saw themselves as defenders of the freedom of scientific inquiry against the possible imposition of a Biblical interpretation of man's creation, rather than as supporters of slavery although their doctrine was seen by many Northerners as an apology for the bondage of Blacks.

653. ———. "The Scientific Approach to the Study of Man in America." *Journal of World History*, 8 (1965): 768-788.

Stresses that the ethic of equality in a land which contained three different races placed the sciences of man in a unique position--questions about the differ-

ences between races had tremendous political and social
significance.

654. Stocking, George W., Jr. "The Boas Plan for the Study of
        American Indian Languages." *Studies in the
        History of Linguistics: Traditions and Paradigms.*
        Edited by Dell Hymes. Bloomington and London:
        Indiana University Press, 1974, pp. 454-484.

Views Boas as a self-taught linguist who wanted
to organize American Indian linguistics according to his
personal perceptions. Rejects Gruber's thesis that
Horatio Hale was a major influence upon Boas (see item 637).

655. ———. "Franz Boas and the Founding of the American
        Anthropological Association." *American
        Anthropologist*, 62 (1960): 1-17.

Examines the dispute between Boas and W.J. McGee
over the structure of the American Anthropological
Association. Argues that Boas wanted an elitist organization in accord with his efforts to professionalize
anthropology, while McGee preferred a broad, inclusive
organization which would encourage local societies and
embrace amateurs.

656. ———. "From Physics to Ethnology: Franz Boas' Arctic
        Expedition as a Problem in the Historiography
        of the Behavioral Sciences." *Journal of the
        History of the Behavioral Sciences*, 1 (1965):
        53-66.

Argues that Boas's realization of the relativity
and arbitrariness of customary behavior was not a sudden
conversion experience, as it has usually been portrayed,
but rather a gradual raising of the sensitivity of a
scientist already interested in man's interaction with
nature and the relationship between the physical and the
psychic and between perception and reality.

ARCHAEOLOGY

657. Brew, J.O. "Introduction." *One Hundred Years of
        Anthropology.* Cambridge, Mass.: Harvard
        University Press, 1970, pp. 5-25.

Presents a brief history of the Peabody Museum
of Archaeology. Rejects earlier theories that the
Peabody remained officially independent of Harvard
University until 1896 because the subject of archaeology

was thought inappropriate for academic recognition. Contends that the Peabody remained independent because of
financial limitations imposed by the Harvard Corporation
on professional and scientific departments.

658. Fagan, Brian. *Elusive Treasures: The Story of Early*
         American Archaeologists in the Americas. New
         York: Charles Scribner's Sons, 1977. xiv +
         369 pp. Bibliography, Index.

    Presents a well-illustrated history of archaeology
in both North and South America, demonstrating the similarities and differences in approach in the two hemispheres.
Finds that both scientific exploration and the raids by
pot hunters increased sharply in the wake of the exhibits
on Indian life in the Southwest at the Philadelphia
Centennial Exposition. Contends that these exhibits
made Americans aware for the first time of the wealth
of their archaeological holdings.

659. Fitting, James E., editor. *The Development of North*
         American Archaeology: Essays in the History of
         Regional Traditions. Garden City, N.Y.: Anchor
         Press, 1973. viii + 309 pp. Index.

    Surveys and summarizes the history of archaeo-
logical ideas. Finds distinct regional traditions.
Argues that anomalies began to challenge the paradigms
of North American archaeology in the early 1960s, leaving
the discipline in a state of crisis.

660. Schuyler, Robert L. "The History of American Archaeology:
         An Examination of Procedure." *American*
         Antiquity, 36 (1971): 383-409.

    Criticizes the lack of contact with concurrent
American social and intellectual history and the failure
to utilize quantitative evidence. Provides case studies
to illustrate the problem.

661. Sheftel, Phoebe Sherman. "The Archaeological Institute of
         America, 1879-1979: A Centennial Review."
         American Journal of Archaeology, 83 (1979):
         3-17.

    Views the Institute as a result of Charles
Eliot Norton's effort to raise the cultural level of
America by encouraging classical studies. Finds that
there was considerable disagreement among the founders
over the issue of restricting activities to classical

archaeology. Discusses the eventual inclusion of Native American sites to the purview of the Institute.

662. Tax, Thomas Gilbert. "The Development of American Archaeology, 1800-1879." Ph.D. dissertation, University of Chicago, 1973.

Studies the field during its transformation from a "dilettante's pastime to an emergent professional science." Argues that Americans inherited the British antiquarian tradition which they subsequently nurtured; this tradition reached maturity in America during the 1840s. Finds a subsequent shift towards empiricism. Concludes that by 1879 antiquarianism was no longer viable, having been replaced by an analytical, professional science.

663. ————. "E. George Squier and the Mounds, 1845-1860." *Towards a Science of Man: Essays in the History of Anthropology*. Edited by Timothy H.H. Thoresen. The Hague and Paris: Mouton Publishers, 1975, pp. 99-124.

Argues that Joseph Henry's insistence on the most rigorous scientific standards and Squier's enthusiasm combined to produce a model of scientific archaeology in the form of Squier and Davis's *Ancient Monuments*, published by the Smithsonian in 1848. Credits this book with marking the beginning of the end of the speculative, antiquarian tradition in American archaeology.

664. Willey, Gordon R., and Jeremy A. Sabloff. *A History of American Archaeology*. San Francisco: W.H. Freeman and Company, 1974. 252 pp. Bibliography, Index.

Provides an intellectual history of the efforts to understand the ancestors of the American Indian and Eskimo. Includes as major themes the impact of European ideas, techniques, and professional models, the relationship of archaeologists with ethnologists and anthropologists and the changing modes of explanation. Claims that armchair speculation was the dominant mode until 1840, when it began to give way to classificatory and geographical distribution studies. In 1914 the central concern changed to chronology. Around 1940 archaeologists became interested in function and environmental context. the "New Archaeology" appeared about 1960.

PSYCHOLOGY

665. Camfield, Thomas M. "The Professionalization of American
              Psychology, 1870-1917." *Journal of the History
              of the Behavioral Sciences*, 9 (1973): 66-75.

        Divides the professionalization process into
   three stages: the initial institutional foundations were
   established between 1870 and 1892; from 1892 through 1904
   there was a conscious effort at professionalization through
   the vehicle of the American Psychological Association;
   after 1904 the chief concern of the psychology community
   was the stature of psychology as a science and a
   profession.

666. Cravens, Hamilton, and John C. Burnham. "Psychology and
              Evolutionary Naturalism in American Thought,
              1890-1940." *American Quarterly*, 23 (1971):
              635-657.

        Approaches Behaviorism in psychology as an
   example of the continuing intellectual tradition of
   evolutionary naturalism--the use of developmental models
   and naturalistic levels of explanation for man and his
   culture. Summarizes attempts to understand human behavior
   in this context, counterarguments from the sociologists,
   and the synthetic thesis that heredity and the social
   environment interact dynamically.

667. Hale, Matthew, Jr. *Human Science and Social Order: Hugo
              Münsterberg and the Origins of Applied
              Psychology*. Philadelphia: Temple University
              Press, 1980. xii + 239 pp. Index.

        Argues that this German-born Harvard professor
   was responsible in large part for laying the groundwork
   for the discipline of applied psychology. Views Münsterberg as an aggressive publicist who offered his science
   as a curative and explanatory tool at a time when
   irrational behavior was on the rise. Describes Münsterberg
   as a social reformer who utilized scientific expertise
   to restore and maintain traditional values.

668. Joncich, Geraldine. *The Sane Positivist: A Biography of
              Edward L. Thorndike*. Middletown: Wesleyan
              University Press, 1968. 634 pp. Bibliography,
              Index.

        Depicts Thorndike's goal as the establishment
   of an experimental, applied science of man in which human

nature would be described in terms of matter and energy. Demonstrates Thorndike's influence in both psychology and pedagogy. Provides a detailed description of Teachers College, Columbia University, around the turn of the twentieth century.

669. Kevles, Daniel J. "Testing the Army's Intelligence: Psychologists and the Military in World War I." *The Journal of American History*, 55 (1968): 565-581.

Discusses the psychological testing conducted by Robert M. Yerkes during World War I. Argues that the program was abolished in 1919 because the Army felt the tests were too academically oriented and impractical, represented a threat to the authority of the officer corps, and were unnecessary during periods of reduced manpower needs.

670. Mandler, Jean Matter, and George Mandler. "The Diaspora of Experimental Psychology: The Gestaltists and Others." *The Intellectual Migration: Europe and America, 1930-1960*. Edited by Donald Fleming and Bernard Bailyn. Cambridge, Mass.: Harvard University Press, 1969, pp. 371-419.

Argues that the intellectual attitudes of the German immigrants were not compatible with the empiricist, comparative, pragmatic American school of experimental psychology. Finds a distinct xenophobic attitude exhibited by American psychologists towards the immigrants. Rejects Wellek's conclusion about the influence of the German immigrants (see item 673) as an overstatement.

671. Pastore, Nicholas. "The Army Intelligence Tests and Walter Lippmann." *Journal of the History of the Behavioral Sciences*, 14 (1978): 316-327.

Deals with Lippmann's recognition that the interpretation of intelligence tests was fraught with political consequences and was dependent upon the political sympathies of the psychologists. Contends that during the period after World War I intellectual elitists wanted to utilize the Army test results as an excuse to restructure American society.

672. Ross, Dorothy. *G. Stanley Hall: The Psychologist as Prophet*. Chicago and London: University of Chicago Press, 1972. xix + 482 pp. Bibliography, Index.

   Treats the life, career, and work of one of the major figures in late nineteenth and early twentieth-century psychology. Argues that he played a vital role in the professionalization of psychology through his research, editorship of the *American Journal of Psychology*, presidency of Clark University, and presidency of the American Psychological Association. Explains Hall's motivations, problem selection, and conclusions, especially in regard to his work on adolescence and children, partly through psychological analysis of the scientist.

673. Wellek, Albert. "The Impact of the German Immigration on the Development of American Psychology." *Journal of the History of the Behavioral Sciences*, 4 (1968): 207-229.

   Identifies members of six schools of psychology, including the Gestalt and Vienna Schools, among the refugees from the Nazis. Argues that this immigration was responsible for the declining influence of Behaviorism in the United States.

SOCIOLOGY

674. Becker, Ernest. "The Tragic Paradox of Albion Small and American Social Science." *The Lost Science of Man*. New York: George Braziller, 1971, pp. 1-70.

   Sees sociology's original goal as reformation based on knowledge of the social process. Contends that specialization and discipline building led to abstraction and the loss of the power to solve human problems. Concludes that Small's attempts to create a discipline of sociology resulted in the destruction of the original dream of sociologists.

675. Bernard, Luther Lee, and Jessie Bernard. *Origins of American Sociology: The Social Science Movement in the United States*. New York: Thomas Y. Crowell Company, 1943. xiv + 866 pp. Indices.

   Argues that the Social Science Movement was an attempt to produce a theory and methodology for the improvement of the human condition. Traces it to the liberal, democratic social theories of eighteenth-century England and France which emphasized reason, natural laws, and the individual. Finds that the movement successively took the form of an utopian aspiration, a metaphysical speculation, an attempt to establish realistic principles of social welfare, an organization for the discussion and promotion of social reforms, and an academic discipline-- which ultimately evolved into the science of sociology.

676. Carneiro, Robert L. "Herbert Spencer's *The Study of Sociology* and the rise of Social Science in America." *Proceedings of the American Philosophical Society*, 118 (1974): 540-554.

Traces the impact of one of the most influential books in its field.

677. Cravens, Hamilton. "The Abandonment of Evolutionary Social Theory in America: The Impact of Academic Professionalization upon American Sociological Theory, 1890-1920." *American Studies*, 12 (1971): 5-20.

Shows that the process of professionalization led sociologists to reject models, metaphors, and determinants explicitly drawn from the evolutionary natural sciences in favor of their own assumptions and analogies.

678. Dibble, Vernon K. *The Legacy of Albion Small*. Chicago and London: The University of Chicago Press, 1975. x + 255 pp. Appendices, Index.

Argues that Small combined a belief in the objectivity of sociology with a commitment to the role of the social scientist as reformer, synthesizing an allegiance to the scientific method with a belief that sociology was an ethical discipline. Links Small's vision of sociology as ethical in nature with his liberal Protestant background and his perception of faculty members as moral leaders. Concludes that his organizational success doomed his intellectual offerings; his efforts to establish sociology as a discipline opened the way for specialists who ignored the sweeping generalizations characteristic of his own work.

679. Diner, Stephen J. "Department and Discipline: The
                Department of Sociology at the University of
                Chicago, 1892-1920." *Minerva*, 13 (1975):
                514-553.

        Examines the early history of the Department of
Sociology of the University of Chicago, where sociology
was formed into an academic subject. Argues that the key
to this department's success was its independence from
other disciplines; in other academic environments,
sociology was subordinated to history, political science,
or anthropology.

680. Faris, Robert E.L. *Chicago Sociology, 1920-1932*. San
                Francisco: Chandler Publishing Company, 1967.
                xiv + 163 pp. Appendices, Index.

        Contends that the Department of Sociology of the
University of Chicago became the dominant American center
for sociology because there was no prescribed doctrine or
authoritative leader; rather, there was concern for
developing a useful methodology and a willingness to
borrow from other schools. Identifies urban ecology
research as the most distinctive development at the
Chicago department. Chicago sociologists used the city
as a laboratory, focusing on intercorrelations, the
natural evolution of the city's zones, and urban behavior.
Argues that Chicago lost its dominant role because of
the improvement in the sociology departments in other
universities, not because of a decline in the Chicago
department.

681. Fine, William F. *Progressive Evolutionism and American
                Sociology, 1890-1920*. Ann Arbor: UMI Research
                Press, 1979. xv + 302 pp. Appendix,
                Bibliography, Index.

        Identifies progressive evolutionism--a compre-
hensive, synthetic view of man, society, and history,
reflecting both the dynamism and progress of man and the
complex, organic nature of the socio-cultural world--as
the prevailing paradigm of early American sociology.
Argues that it represented an attempt to construct an
interpretation of humans which would both legitimatize
the discipline of sociology and support the Progressive
political and social reforms.

682. Fuhrman, E.R. "Images of the Discipline in Early American Sociology." *Journal of the History of Sociology*, 1 (1978): 91-116.

Analyzes the image of the discipline in the work of six early American sociologists. Finds that most thought the natural sciences were the examplars.

683. Heyl, John D., and Barbara S. Heyl. "The Sumner-Porter Controversy at Yale: Pre-Paradigmatic Sociology and Institutional Crisis." *Sociological Inquiry*, 46 (1976): 41-49.

Rejects prior interpretations that this was a clash between science and religion. Describes it as a struggle by a discipline still in its pre-paradigmatic stage to achieve academic recognition.

684. Hinkle, Roscoe C. *Founding Theory of American Sociology, 1881-1915*. Boston, London, and Henley: Routledge and Kegan Paul, 1980. xiv + 376 pp. Bibliography, Index.

Advocates periodizing the history of sociology according to the dominant or ascending theoretical concepts as demonstrated by the published literature. Analyzes the work of six early leaders in American sociology and concludes that the founding theory was more substantive-ontological--concerned with the nature of social phenomena --than epistemological-methodological--concerned with the nature and source of social knowledge.

685. Martindale, Don. "American Sociology Before World War II." *Annual Review of Sociology*, 2 (1976): 121-143.

Argues that the differences in the structure of American society led American sociologists to reject the European concept of society as a superindividual organism (collectivism) in favor of an "unreflective individualism."

686. Matthews, Fred H. *Quest for an American Sociology: Robert E. Park and the Chicago School*. Montreal and London: McGill-Queen's University Press, 1977. ix + 278 pp. Bibliography, Index.

Describes Park as a major factor in the trans-formation of sociology from a mixture of speculation about the nature of society, Christian philanthropy, and descriptive studies of social problems to a powerful

academic discipline. Traces Park's career from journalism
to academia to press relations for Booker T. Washington
and finally back to academia, when, at age 49, he joined
the faculty of the University of Chicago in 1913. Credits
Park with weaning scholars away from racial typing, re-
defining race as a subjective category, and treating
racial prejudice as a function of group and status
conflict, rather than an automatic inherent reaction,
resulting in a new framework for understanding racial
prejudice.

687. Schwendinger, Herman, and Julia R. Schwendinger. *The
          Sociologists of the Chair: A Radical Analysis
          of the Formative Years of North American
          Sociology (1883-1922)*. New York: Basic Books,
          1974. xxviii + 609 pp. Bibliography, Index.

Utilizes explicitly Marxist analysis in contending that there was a reciprocal relationship between
academic sociology and the corporate version of the liberal
ideology which evolved during this period. Views the
writings of leading American sociologists as rationales
for capitalism, imperialism, racism, and sexism. Dis-
misses distinctions and controversies within the
sociological establishment as insignificant.

688. Sutherland, David Earl. "Who Now Reads European Sociology?
          Reflections on the Relationship Between European
          and American Sociology." *Journal of the History
          of Sociology*, 1 (1978): 35-53.

Argues that earlier studies have overemphasized
the intellectual influence of European sociology upon
the development of American sociology, while also
perpetuating the myths that European sociology was not
empirical and that empiricism was the key factor in the
rise of American sociology to a position of hegemony.
Claims that American superiority was due to the depart-
mental structure of its universities, in contrast to
the European practice of a single chair in a field.

CHAPTER VI: TECHNOLOGY AND AGRICULTURE

INSTRUMENTS AND INSTRUMENT-MAKERS

689. Bedini, Silvio A. "Artisans in Wood: The Mathematical
Instrument-Makers." *America's Wooden Age:
Aspects of its Early Technology.* Edited by
Brooke Hindle. Tarrytown: Sleepy Hollow
Restorations, 1975, pp. 85-119.

Stresses that making instruments was relatively
easy; it was the graduation and inscription of scales
which was difficult. Observes that there were relatively
few instrument-makers because other forms of artisanship
offered better economic opportunities.

690. ————. *Early American Scientific Instruments and Their
Makers.* United States National Bulletin 231.
Washington, D.C.: Smithsonian Institution
Press, 1964. xii + 184 pp. Appendix,
Bibliography, Index.

Discusses only mathematical instruments designed
for practical uses, such as surveying or navigational
instruments, ignoring philosophical apparatus used in
the laboratory or classroom. Identifies approximately
one hundred active instrument-makers in eighteenth-century
America. Provides biographical information for them.
Finds that they often acquired their skills through
apprenticeship. Lists the material in the collections
of the Smithsonian. Includes many illustrations.

691. ————. *Thinkers and Tinkers: Early American Men of
Science.* New York: Charles Scribner's Sons,
1975. xix + 520 pp. Glossary, Bibliography,
Index.

Describes the activities of mathematical
practitioners--a group which included mapmakers, surveyors,
navigators, instrument-makers, and a few men with

international scientific reputations like Franklin and
Rittenhouse--from the seventeenth century through the
first third of the nineteenth. Provides many illustrations
of instruments and a glossary of technical terms.

692. Cassedy, James H. "Applied Microscopy and American Pork
        Diplomacy: Charles Wardell Stiles in Germany,
        1898-1899." *Isis*, 62 (1971): 5-20.

        Discusses the role of Charles W. Stiles in an
American challenge to the German biomedical community
involving the inspection of American pork exported to
Germany. Illuminates the development of the position of
scientific attaché in American embassies and the growing
internationalism in American science.

693. ————. "The Microscope in American Medical Science,
        1840-1860." *Isis*, 67 (1976): 76-97.

        Traces the introduction and diffusion of achro-
matic microscopes--essential for microscopic research in
the biological sciences--in the United States.

694. Eddy, John A. "The Schaeberle 40-ft. Eclipse Camera of
        Lick Observatory." *Journal for the History of
        Astronomy*, 2 (1971): 1-22.

        Describes the camera which produced unequalled
photographs of the solar corona during fourteen eclipses.

695. ————. "Thomas A. Edison and Infra-red Astronomy."
        Journal for the History of Astronomy, 3 (1972):
        165-187.

        Describes the development of the tasimeter and
its use at the 1878 eclipse. Argues that Edison's
invention was too delicate and erratic for practical
astronomical use.

696. Hughes, Thomas Parke. *Science and Instrument-maker:
        Michelson, Sperry, and the Speed of Light*.
        Smithsonian Studies in History and Technology,
        Number 37. Washington, D.C.: Smithsonian
        Institution Press, 1976. iii + 18 pp.

        Presents a case study of the dependence of a
scientist upon the mechanical ingenuity of an instrument-
maker.

697. Mills, Deborah J. "George Willis Ritchey and the Development of Celestial Photography." *American
         Scientist*, 54 (1966): 64-94.

    Credits Ritchey with making the large photographic reflector a basic tool of twentieth-century
astronomical research through his perfection of optical,
mechanical, and photographic parts.

698. Padgitt, Donald L. *A Short History of the Early American
         Microscopes*. The Microscope Series, Volume 12.
         London and Chicago: Microscope Publications,
         Ltd., 1975. xi + 147 pp. Index.

    Concentrates on the improvements in the
mechanical design of microscopes built in the United
States during the nineteenth century. Demonstrates the
dominant role of immigrant craftsmen. Includes short
biographical sketches of makers, designers, and dealers.

699. Smart, Charles E. *The Makers of Surveying Instruments
         in America Since 1700*. 2 Volumes. Troy: Regal
         Art Press, 1962-1967. xxvi + 266 pp.
         Bibliography, Index.

    Describes surveying instruments. Supplies
brief histories of the companies or individuals who
manufactured them. Avoids qualitative evaluations of
the skill or significance of the manufacturers.

700. Warner, Deborah Jean. *Alvan Clark & Sons: Artists in
         Optics*. United States National Museum Bulletin
         274. Washington, D.C.: Smithsonian Institution
         Press, 1968. 120 pp.

    Includes brief biographical sketches of the
Clarks and a catalogue of all known Clark instruments.

701. ————. "The American Photographical Society and the
         Early History of Astronomical Photography in
         America." *Photographic Science and Engineering*,
         11 (1967): 342-347.

    Finds a number of pioneer astrophotographers
among the membership of the first photographic society
in the United States. Argues that the society was a
source of technical information.

702. ———— "Lewis M. Rutherfurd: Pioneer Astronomical
          Photographer and Spectroscopist." *Technology
          and Culture*, 12 (1971): 190-216.

     Illustrates the role of technical improvements
in instrumentation in advancing science. Focuses on the
design and construction of apparatus vital to the development of two new disciplines. Contains a plea to historians
to pay more attention to the instrument-makers, the
"producers of works rather than words."

703. Willard, Berton C. *Russell W. Porter: Arctic Explorer,
          Artist, Telescope Maker*. Freeport, Maine:
          Bond Wheelwright Company, 1976. xiii + 274 pp.
          Bibliography.

     Offers insights into the development of amateur
telescope making in the twentieth century. Discusses the
design problems associated with the 200-inch telescope.

SCIENCE-TECHNOLOGY RELATIONSHIP

704. Dunlap, Thomas R. "Science as a Guide in Regulating
          Technology: The Case of DDT in the United
          States." *Social Studies of Science*, 8 (1978):
          265-285.

     Demonstrates that during most of the period of
DDT use, neither policy makers nor the general public
had the technical information necessary to make rational
decisions. Blames this failure on the lack of analytical
tools to trace the pesticide through the environment,
the organization and funding of science along disciplinary
lines, and the lack of adequate guides to research.

705. Kogan, Herman. *A Continuing Marvel: The Story of the
          Museum of Science and Industry*. Garden City:
          Doubleday and Company, 1973. iii + 233 pp.
          Appendix, Bibliography, Index.

     Traces the evolution of the museum from its
original conception as an American version of the
Deutsches Museum—a historically oriented museum with
hands-on working models—to the present format, developed
by Lenox Lohr (appointed president of the museum in 1940),
of a mass education medium emphasizing contemporary
accomplishments. Credits Lohr with the idea of having
industry design, build, and pay for exhibits, limiting
the function of the museum staff to insuring that no
direct advertising took place. Contains little inter-

pretation and no documentation but claims to have been
based on the manuscript holdings of the museum.

706. Layton, Edwin T., Jr. "American Ideologies of Science
     and Engineering." *Technology and Culture*, 17
     (1968): 688-701.

     Analyzes three ideologies which represent the
spectrum of the possible connections between engineering
artifacts and the ideal world of theoretical physics,
ranging from the parasitic dependence of technological
progress upon basic science to the symbiotic relationship
between two distinct subcultures with differing objectives.

707. ————. "Millwrights and Engineers, Science, Social
     Roles, and the Evolution of the Turbine in
     America." *The Dynamics of Science and Tech-
     nology: Social Values, Technical Norms, and
     Scientific Criteria in the Development of
     Knowledge*. Edited by Wolfgang Krohn, Edwin
     T. Layton, Jr., and Peter Weingart. Dordrecht:
     Reidel, 1978, pp. 61-87.

     Examines the parallel research on turbines by
French engineers and American wheelwrights during the
early nineteenth century. Finds that the differences
reflected disparities in their respective scientific
traditions. Concludes that American hydraulic engineers
were a marriage of the two traditions.

708. ————. "Mirror-Image Twins: The Communities of Science
     and Technology." *Nineteenth-Century American
     Science: A Reappraisal* (item 32)

     Argues that during the nineteenth century
American technology underwent a "scientific revolution,"
adopting the methodology of science and discarding its
craft traditions. Describes the resulting community as
quite distinct from the scientific community, with a
different objective (scientists want to know; technologists want to do).

709. ————. "Scientists and Engineers: The Evolution of the
     IRE." *Proceedings of the IEEE*, 64 (1976):
     1390-1392.

     Relates the history of the Institute of Radio
Engineers, the most scientifically oriented and democratic
American engineering society. Attributes these characteristics to the highly scientific nature of the technology
involved.

710. Molella, Arthur P. "The Electric Motor, the Telegraph,
                 and Joseph Henry's Theory of Technological
                 Progress." *Proceedings of the IEEE*, 64
                 (1976): 1273-1278.

          Contrasts Henry's support of the development of
     the telegraph with his rejection of the electric motor as
     impractical. Demonstrates that Henry's attitude resulted
     from his belief that technological progress was dependent
     upon the state of scientific knowledge.

711. ———, and Nathan Reingold. "Theorists and Ingenious
                 Mechanics: Joseph Henry Defines Science."
                 Science Studies, 3 (1973): 323-351.

          Analyzes Henry's view that the relationship
     between science and technology was hierarchical; it was
     most important to develop a theoretical understanding
     of nature which then could be reduced to practice. Places
     this attitude within the context of the current histori-
     ography of this relationship.

712. Reingold, Nathan. "Alexander Dallas Bache: Science and
                 Technology in the American Idiom." *Technology
                 and Culture*, 11 (1970): 163-177.

          Argues that both the career and attitudes of
     Bache, a key, representative figure in the development of
     American science, reflected the close relationship between
     basic and applied research in this country.

INDUSTRIAL RESEARCH

713. Birr, Kendall. "Industrial Research Laboratories." *The
                 Sciences in the American Context: New Perspec-
                 tives* (item 84), pp. 193-207.

          Surveys the development of the industrial
     research laboratory. Identifies three pre-conditions
     for the rise of these laboratories: industrial leaders
     who felt a need for scientific and technological assistance;
     large corporations with the financial resources to support
     industrial research; and successful models to follow, such
     as the laboratories in German industries.

714. ———. *Pioneering in Industrial Research: The Story of
                 the General Electric Research Laboratory*.
                 Washington, D.C.: Public Affairs Press, 1957.
                 vii + 204 pp. Index.

Traces the history of the laboratory from its
founding in 1901 until the post-World War II reorganization. Argues that the laboratory was atypical of American
industrial research facilities; it was both exceptionally
successful and unique in some of its practices. Attributes
its success to the support of the mother company, the
quality of its leadership and personnel, the willingness
of GE to conduct fundamental research to expand the
limits of scientific knowledge in the field of electricity,
and the ability of GE to exploit commercially the discoveries of the laboratory.

715. ————. "Science in American Industry." *Science and
Society in the United States* (item 99),
pp. 35-80.

Contends that science and technology were
separate and different enterprises which did not begin
interacting until the mid-nineteenth century. Finds only
two industries prior to World War I--the electrical and
the chemical--where science had information of value to
industry and the industry was willing to apply that
information.

716. Brittain, James E. "C.P. Steinmetz and E.F.W. Alexander-
son: Creative Engineering in a Corporate
Setting." *Proceedings of the IEEE*, 64 (1976):
1413-1417.

Views the Consulting Engineering Department,
founded at General Electric in 1910, as both an attempt
to institutionalize creative engineering and a challenge
to the concept of engineering as a parasite upon science.

717. ————. "The Introduction of the Loading Coil: George
A. Campbell and Michael I. Pupin." *Technology
and Culture*, 11 (1970): 36-57.

Questions the validity of the decision granting
the patent to Pupin. Credits Campbell with the theoretical
analysis which led Bell Telephone to develop the innovation.

718. Fagen, M.D., editor. *A History of Engineering and Science
in the Bell System*. 2 Volumes. Murray Hill:
Bell Telephone Laboratories, 1975-1978. xiii +
1,073 pp., xv + 757 pp. Bibliography, Index.

Examines the role of Bell Laboratories in the
history of communication and the application of science
and engineering to the national defense. Provides

detailed information but little analysis. The first
volume focuses on civilian needs, especially the telephone
and non-voice communications, and on basic research during
the period 1875-1925. Volume two skips over the years
1925-1936 to concentrate on contributions to radar,
communications systems, fire control, air defense, and
other military needs during World War II and the Cold
War; the volume terminates in 1975. The contributors
to the volumes were the technical staff at the Bell
Laboratories, not historians.

719. Hall, Courtney Robert. *History of American Industrial
             Science*. New York: Library Publishers, 1954.
             Reprint. New York: Arno Press, 1972. xix +
             453 pp. Bibliography, Index.

        Concentrates on twentieth-century developments
in such industries as transportation, chemicals, mining
and metallurgy, petroleum, rubber, pulp and paper, food
and clothing production and distribution, and precision
instruments. Fails to define industrial science, but the
term appears to embrace all uses of science for research
and development as well as the impact of regulation and
business organization on such use.

720. Hughes, Thomas P. "Edison's Method." *Technology at the
             Turning Point*. Edited by William B. Pickett.
             San Francisco: San Francisco Press, 1977,
             pp. 5-22.

        Identifies Edison's method as the invention of
entire systems which satisfied predetermined economic
needs, utilizing team research. Contends that Edison
was the leader of a small-scale industrial laboratory.
Questions the historiography which credits General
Electric with pioneering industrial research.

721. Layton, Edwin T., Jr. "Scientific Technology, 1845-1900:
             The Hydraulic Turbine and the Origins of
             American Industrial Research." *Technology
             and Culture*, 20 (1979): 64-89.

        Presents a four-stage case study more represen-
tative of the interaction of science and technology in
the United States than the traditional example of the
chemical industry. Argues that the major difference
between science and technology was evident in the
technologists' skepticism of mathematical models;
science seeks generality at the price of idealization
while technology serves practice at the cost of generality.

722. Miller, John Anderson. *Workshop of Engineers: The Story of the General Engineering Laboratory of the General Electric Company, 1895-1952.* Schenectady: General Electric Company, 1953. 173 pp. Index.

   Presents a house history with neither documentation nor analysis. Represents a genre to be avoided.

723. Pursell, Carroll. "Science and Industry." *Nineteenth Century American Science: A Reappraisal* (item 32), pp. 231-248.

   Contends that the entrepreneurial spirit in America led parts of the very diverse scientific community to worry about applications of science to industrial problems.

724. Rae, John. "The Application of Science to Industry." *The Organization of Knowledge in Modern America, 1860-1920.* Edited by Alexandra Oleson and John Voss. Baltimore and London: The Johns Hopkins University Press, 1979, pp. 249-268.

   Identifies four types of research laboratories applying science to industrial problems: Edison's, representative of private laboratories studying specific problems to generate profitable discoveries; consulting laboratories; nonprofit research institutes; company-owned laboratories.

725. Reich, Leonard S. "Research, Patents, and the Struggle to Control Radio: A Study of Big Business and the Uses of Industrial Research." *Business History Review*, 51 (1977): 208-235.

   Contends that companies often used their research-generated patents to gain or maintain monopoly positions without actually utilizing the innovations. Offers three examples of the "non-productive" use of patents: the prevention of competition, the forestalling of competitors' attempts to gain strong patent positions, and the trading of patents.

726. ―――. "Science." *Encyclopedia of American Economic History.* Edited by Glenn Porter. New York: Charles Scribner's Sons, 1980, 1:281-293.

   Surveys the role of science and scientific methodology in the development of American industry. Argues that science and engineering have been closely linked.

727. Weiner, Charles. "How the Transistor Emerged." *IEEE Spectrum*, 10 (1973): 24-33.

        Argues that "social inventions" played a vital role in the discovery of the transistor, including international fellowships, which aided the development of the quantum and solid state physics which formed the fundamental knowledge behind the transistor; the specialized review publication *Bell System Technical Journal*, which provided a link between academic and industrial physics; and the interdisciplinary mode of organizing research to facilitate communication among specialists and the application of basic science.

728. Wise, George. "A New Role for Professional Scientists in Industry: Industrial Research at General Electric, 1900-1916." *Technology and Culture*, 21 (1980): 408-429.

        Argues that industrial research represented an alternative career for scientists, appealing to those individuals either uncomfortable in academia or unwilling to become entrepreneurs, yet desiring a career involving research.

AGRICULTURE

729. Beardsley, Edward H. *Harry L. Russell and Agricultural Science in Wisconsin*. Madison, Milwaukee, and London: The University of Wisconsin Press, 1969. x + 237 pp. Bibliography, Index.

        Examines the life of one of the key figures in the application of science to agriculture. Describes his research in plant bacteriology, bovine tuberculosis, and cheese curing, pointing out that some of his solutions were not popular with farmers. Contends that during his term as dean of the College of Agriculture of the University of Wisconsin (1907-1930), the College became a model for agricultural research and extension services. Argues that his attitude towards the farm differed from most of his colleagues; he saw agriculture only in economic terms, having no sympathy for the romanticizing of farm life.

730. Fitzharris, Joseph C. "Science for the Farmer: The
         Development of the Minnesota Agricultural
         Experiment Station, 1868-1910." *Agricultural
         History*, 48 (1974): 202-214.

     Contends that the personal interaction between
Station staff, students at the College of Agriculture,
and farmers, combined with the gradual introduction of
scientific methods, enabled the Station to establish a
strong political base, overcome farmer resistance, and
increase productivity.

731. Hilliard, Sam B. "The Dynamics of Power: Recent Trends
         in Mechanization on the American Farm."
         Technology and Culture, 13 (1972): 1-24.

     Traces and analyzes the trend toward smaller
farm machinery during the years between the World Wars,
and the reversal of that trend by the end of World War
II. Credits the decline in the number of farms and the
increase in average farm size for the embracing of
larger tractors and more self-propelled machinery.

732. Rasmussen, Wayne D. "Advances in American Agriculture:
         The Mechanical Harvester as a Case Study."
         Technology and Culture, 9 (1968): 531-543.

     Demonstrates that modern technological advances
in agriculture occur through the adoption of an entire
package of techniques, tools, and plants, rather than
the development of a single innovation.

733. Rosenberg, Charles E. "Rationalization and Reality in
         Shaping American Agricultural Research, 1875-
         1914." *The Sciences in the American Context:
         New Perspectives* (item 84), pp. 143-163.

     Blames the inconsistencies and contradictions
in the designs of the agricultural scientists and the
consequences of their activities--they had hoped their
research would foster the small farmer, but instead it
led to his decline by aiding agribusiness--on the
consistencies between their social assumptions, their
professional needs, and the cultural context of their
activities.

734. Rossiter, Margaret W. *The Emergence of Agricultural*
     Science: Justus Liebig and the Americans,
     1840-1880. New Haven: Yale University Press,
     1975. xiv + 275 pp. Bibliography, Index.

     Examines the efforts to import both the ideas
and the institutional structure of German agricultural
science into the United States, focusing on the careers
of two of Liebig's American students--Eben Horsford and
Samuel W. Johnson--and one of his American critics--John
P. Norton. Argues that there was an initial receptivity
to Liebig's theories, resulting in a successful, if
selective, importation of his ideas. Finds, however,
that Horsford's efforts to duplicate, at Harvard, Liebig's
research laboratory failed because of the high costs.
Traces the evolution of agricultural science at Yale
under Norton and Johnson, which culminated in the establishment of America's first agricultural research station
(1875).

735. ————. *A List of References for the History of Agricul-*
     tural Sciences in America. Davis: Agricultural
     History Center, University of California, 1980.
     viii + 62 pp. Index.

     Provides divisions by scientific field, an
author index, but no annotations. Quite extensive.

736. ————. "The Organization of Agricultural Improvement
     in the United States, 1785-1865." *The Pursuit*
     of Knowledge in the Early American Republic:
     American Scientific and Learned Societies
     from Colonial Times to the Civil War (item
     209), pp. 279-298.

     Shows the evolution of agricultural societies
from gentlemen's clubs run by the local elite to lobbying
organizations for the working farmers. Claims that prior
to 1875, demand side of agricultural improvement (diffusion of knowledge) took precedent over the supply
side (research).

737. ————. "The Organization of the Agricultural Sciences."
     The Organization of Knowledge in Modern America,
     1860-1920. Edited by Alexandra Oleson and John
     Voss. Baltimore and London: The Johns Hopkins
     University Press, 1979, pp. 211-248.

     Describes this as a transitional period in
American agriculture; the diminution of the natural

fertility was forcing farmers to become more dependent upon science. Finds that the agricultural sciences were becoming more professionalized and specialized. Surveys activities in six subfields and discovers differential growth.

738. Schlebecker, John T. *Bibliography of Books and Pamphlets on the History of Agriculture in the United States, 1607-1967.* Santa Barbara: Clio Press, 1969. vii + 183 pp. Index.

> Includes, in addition to the usually secondary works, autobiographies, fiction, and historically oriented primary sources. The subject index is fairly complete.

739. ———. *Whereby We Thrive: A History of American Farming, 1607-1972.* Ames: The Iowa State University Press, 1975. x + 342 pp. Bibliography, Index.

> Surveys developments in land policy, marketing, farm technology, and the impact of science upon agriculture. Stresses the integrated nature of mechanization and technological improvement--isolated advances made little economic sense because production was only as rapid and efficient as its slowest and least efficient element.

740. Sharrer, G. Terry. *1001 References for the History of American Food Technology.* Davis: Agricultural History Center, University of California, 1978. 103 pp. Index.

> Includes very selective and brief annotations. Provides a scattering of qualitative evaluations, especially of a positive nature. Restricts primary source entries to the "first books in their field published in America."

741. Stephens, Lester D. "Farish Furman's Formula: Scientific Farming and the 'New South'." *Agricultural History*, 50 (1976): 377-390.

> Offers Furman as an example of the popularizer of a careful and narrow application of science to farming--in his case the preparation of a specific fertilizer for cotton in the wake of a chemical analysis of the plant.

742. Walsh, Thomas R. "Charles E. Bessey and the Transformation
          of the Industrial College." *Nebraska History*,
          52 (1971): 383-409.

          Describes the transformation of a typical
     agricultural and mechanics college, with low admission
     standards and work-study programs, into a center for both
     basic and applied scientific research. Fails, however,
     to explain how Bessey accomplished this.

743. Whitehead, Vivian B. *A List of References for the History
          of Agricultural Technology*. Davis: Agricultural
          History Center, University of California, 1979.
          vii + 76 pp. Index.

          Emphasizes agricultural machinery. Includes
     descriptive annotations.

744. Wik, Reynold M. "Benjamin Holt and the Invention of the
          Track-Type Tractor." *Technolgy and Culture*,
          20 (1979): 90-107.

          Focuses on the contributions of a man who
     exhibited the flexibility to improve and redesign existing
     technology to meet changing needs.

745. ————. "Henry Ford's Science and Technology for Rural
          America." *Technology and Culture*, 3 (1962):
          247-258.

          Contends that Ford's greatest contribution to
     American agriculture was his effort to ensure the super-
     sedure of animal power by mechanical power. Includes
     examples of Ford's support of the application of scientific
     technology to agriculture.

746. ————. "Science and American Agriculture." *Science and
          Society in the United States* (item 99),
          pp. 81-106.

          Discusses the various means of disseminating
     agricultural information, the essential role of govern-
     ment support on all levels, and the impact of mechanical
     technology.

747. ————. *Steam Power on the American Farm*. Philadelphia:
          University of Pennsylvania Press, 1953. xi +
          288 pp. Appendix, Bibliography, Index.

          Analyzes the role of steam power, the first
     successful mechanical prime mover used for agricultural

purposes, in meeting the power crisis in the major
grain-growing regions of the United States and in the
conquest of the agricultural frontiers in the West.
Argues that horsepower had proved inadequate for large-
scale threshing and plowing.

ENGINEERING

748. American Society of Civil Engineers. *A Biographical
Dictionary of American Civil Engineers*. ASCE
Historical Publication No. 2. New York:
American Society of Civil Engineers, 1972.
x + 163 pp.

Contains 170 entries for significant civil
engineers born before the Civil War. Includes a list of
the 2,500 civil engineers in the Biographical Archives
of American Civil Engineers at the Smithsonian Institution.

749. Brittain, James E., and Robert C. McMath, Jr. "Engineers
and the New South Creed: The Formation and
Early Development of Georgia Tech." *Technology
and Culture*, 18 (1977): 175-201.

Argues that Georgia Tech was the result of the
convergence of the New South Creed--the demands for an
industrialized South--and changes in professional
engineering education. Relates how Georgia Tech represented a victory for the school culture over the shop
culture.

750. Calhoun, Daniel Hovey. *The American Civil Engineer:
Origins and Conflict*. Cambridge, Mass.: The
Technology Press, 1960. xiv + 295 pp.
Appendices, Bibliography, Index.

Concentrates on the civil engineers working on
major antebellum internal improvement projects. Provides
a case study of the development of the role of the
salaried professional, loyal to both an organization and
a set of professional standards. Presents various
examples of career patterns which demonstrate this
development. Shows that the depressed economic conditions
of the late 1830s and early 1840s were an important factor
in raising the professional consciousness of the engineers.

751. Calvert, Monte A. *The Mechanical Engineer in America,*
     1830-1910: Professional Cultures in Conflict.
     Baltimore: The Johns Hopkins Press, 1967.
     xviii + 296 pp. Bibliography, Index.

       Describes two cultures within the developing
     mechanical engineering profession. The older "shop
     culture" was composed of shop-trained engineers, often
     of high social class, who were both technicians and
     entrepreneurs, and had status both as engineers and
     businessmen. The newer "school culture" consisted of
     college-educated corporate employees whose social status
     was derived solely from their professional standing.
     Explains aspects of the history of engineering education,
     professional associations, and attempts at standardization
     in terms of the conflict between the two cultures.

752. Floud, R.C. "The Adolescence of American Engineering
           Competition, 1860-1900." *The Economic History*
           Review, 2nd Series, 27 (1974): 57-71.

       Rejects the theory that the British ignored
     American innovations, until the cycle boom of the 1890s
     brought home the undeniable value of those innovations,
     because of engineering conservatism. Offers the alterna-
     tive explanation that the Americans were ignored because
     they were not economically competitive. Views the 1890s
     boom as a response to the increased efficiency of American
     manufacturing practices, resulting in the increased
     competitiveness of American engineering goods.

753. Layton, Edwin T., Jr. *The Revolt of the Engineers: Social*
           Responsibility and the American Engineering
           Profession. Cleveland and London: The Press
           of Case Western Reserve University, 1971.
           xvi + 286 pp. Bibliography, Index.

       Traces the history of the engineering profession
     as it attempted to reconcile a fundamental dichotomy: an
     engineer was both a professional, which implied autonomy,
     collegial control of profession work, and a sense of
     social responsibility, and an employee of a profit-making
     business which demanded loyalty to the organization.
     Discusses the activities during the first quarter of the
     twentieth century of a small band of engineers who felt
     their profession had a special obligation to ensure that
     technological change resulted in benefits for the human
     race. Finds, however, that disillusionment with the
     application of science to human problems, the growing

domination of the engineering field by business interests,
and, ultimately, a preoccupation with their own employment problems during the Depression ended such activities.

754. ————. "Science, Business, and the American Engineer."
The Engineer and the Social System. Edited by
Robert Perrucci and Joel E. Gerstl. New York,
London, Sydney, and Toronto: John Wiley & Sons,
1969, pp. 51-72.

Assesses the conflicting roles of the engineer
as loyal employee of a business organization which wishes
to test his knowledge in the marketplace and as scientific
professional interested only in expanding knowledge.
Judges most specialized engineering societies to be
trade associations rather than professional groups.

755. ————. "Veblen and the Engineers." *American Quarterly*,
14 (1962): 62-72.

Credits Veblen's belief that engineers would
constitute the revolutionary class in America to his
misreading of the early twentieth-century revolt of
reform-minded engineers (see item 753) as a rejection
of the business culture and his previous conclusion that
the engineer was the personification of the creative
instinct, whereas the businessman personified the anti-
thesis--the predatory instinct.

756. McMahon, A. Michal. "Corporate Technology: The Social
Origins of the American Institute of Electrical
Engineers." *Proceedings of the IEEE*, 64 (1976):
1383-1390.

Finds that the AIEE was created by corporate
managers and self-taught inventor-manufacturers, primarily
from the telegraph industry. Discovers that this
generation was quickly displaced by academically trained
engineers employed by the large electrical corporations.

757. Merritt, Raymond H. *Engineering in American Society,
1850-1875.* Lexington: The University Press of
Kentucky, 1969. xi + 199 pp. Bibliography,
Index.

Defines engineers as functional intellectuals
who obtained professional status through the successful
application of knowledge to problems. Argues that the
engineers provided the leadership in the transformation
of American life by technology. Discusses the role of

the engineer as the first professional executive (men
trained to organize and lead complex institutions) and
their ascendency over the lay managers in American
industry (entrepreneurs, political leaders, investors).

758. Morgan, Arthur E. *Dams and Other Disasters: A Century of
     the Army Corps of Engineers in Civil Works.*
     Boston: Porter Sargent, 1971. xxv + 422 pp.
     Index.

     Represents a diatribe against the Corps and
     West Point. Although historical data is occasionally
     utilized to make a point, the author depends primarily
     upon his own personal experiences. Attacks the Corps
     for conservatism and a lack of respect for civilian
     authority.

759. Noble, David F. *America by Design: Science, Technology,
     and the Rise of Corporate Capitalism.* New
     York: Alfred A. Knopf, 1977. xxvi + 384 pp.
     Index.

     Argues that American engineers consciously
     subordinated technology to corporate interests, designing
     a new social order dominated by the private corporation.
     Finds that the results were the endurance of a static or
     even outmoded capitalist system and the prevention of the
     social revolution implied by the modes of production
     possible through scientific technology. Investigates
     the activities of these engineers in standardization,
     patent reform, the organization of research, and the
     transformation of higher education. Presents an explicit-
     ly Marxist interpretation.

760. Pursell, Carroll W., Jr. "The Technical Society of the
     Pacific Coast, 1884-1914." *Technology and
     Culture*, 17 (1976): 702-717.

     Describes the focal point for the professional
     aspirations and activities of West Coast engineers prior
     to the establishment of strong ties with the national
     professional and scientific organizations.

761. Reynolds, Terry S. "American Engineering and British
     Technical Observers: The First Two Hundred
     Years." *Transactions of the Wisconsin Academy
     of Sciences, Arts, and Letters*, 64 (1976):
     83-108.

Finds that British observers thought Americans sacrificed safety, aesthetic appeal, strength, and durability for cheap and rapid construction. The British were impressed, however, by American methods of mass production, which enabled them to assemble parts into the whole, while Europeans had to fit parts into a whole.

762. Rouse, Hunter. *Hydraulics in the United States, 1776-1976.* Iowa City: Institute of Hydraulic Research of The University of Iowa, 1976. ix + 238 pp. Index.

Lists important contributors, events, and publications but often lacks analysis, especially of the scientific content and the social and intellectual factors. The major exception to this criticism is the analysis of John R. Freeman's influence on federal hydraulics during the first third of the twentieth century, which led to the establishment of the National Hydraulic Laboratory over the opposition of the Corps of Engineers. Contends that the hydraulic engineers did not respect the work of the Corps.

763. Roysdon, Christine, and Linda A. Khatri. *American Engineers of the Nineteenth Century: A Biographical Index.* New York and London: Garland Publishing, 1978, xv + 247 pp.

Provides citations to obituary notices and biographical accounts of approximately two thousand engineers who died during the nineteenth century.

764. Servos, John W. "The Industrial Relations of Science: Chemical Engineering at MIT, 1900-1939." *Isis*, 71 (1980): 531-549.

Responds to Noble's analysis of academic relations with industry (see item 759) by studying MIT's attempts to tap the financial resources of the business community. Finds that the chemical engineers eventually rejected the concept that industry could provide a stable base for research in an academic setting because of the restrictions on research and publication imposed by the contributors.

765. Sinclair, Bruce. *A Centennial History of the American Society of Mechanical Engineers, 1880-1980.* Toronto, Buffalo, and London: University of Toronto Press, 1980. xii + 256 pp. Appendix, Index.

Views the founding of the ASME as part of larger institutional building process of the late nineteenth century. Maintains that the essence of the history of the ASME has been the interplay of its two basic objectives: social--including both intrasocietal and non-technical (e.g., political lobbying) extrasocietal relations--and technical. Discusses the conflict between the New York City headquarters, more concerned with the social, and the national membership, concerned with the technical. Analyzes the activities of the ASME in standardization, arguing that its members' motivations were a combination of aesthetics, economics, and the sense of professional responsibility.

766. Spence, Clark C. *Mining Engineers & the American West: The Lace-Boot Brigade, 1849-1933.* New Haven and London: Yale University Press, 1970. xii + 407 pp. Bibliography, Index.

Presents these engineers as well-trained, versatile professionals from middle-class backgrounds who helped spread American technology throughout the world. Depends more upon qualitative impressions than quantitative data for evidence. Provides descriptions of their activities as mine managers or self-employed consultants.

767. Terman, Frederick E. "A Brief History of Electrical Engineering Education." *Proceedings of the IEEE*, 64 (1976): 1399-1407.

Finds that the curriculum changed in response to changes in electrical technology: the first programs in the 1880s met the needs of electrical manufacturers; communications entered the curriculum in the wake of the post World War I growth of radio; more theory was added after World War II to prepare students for the new electronics developed during the war.

768. Wisely, William H. *The American Civil Engineer, 1852-*
          1974: The History, Traditions and Developments
          of the American Society of Civil Engineers.
          New York: American Society of Civil Engineers,
          1974. ix + 464 pp. Appendices, Index.

     Discusses the history of civil engineers under
the rubrics of professionalization, civil engineering,
and public service. Avoids discussions of living
engineers, which leads to a view of the society as an
entity with a life of its own, rather than an organization
which reflected the collective goals of a group of human
beings; the result is a history without discussions of
motivation or internal societal politics. Includes a
considerable amount of useful data in the appendices.

MASS PRODUCTION

769. Battison, Edwin A. "Eli Whitney and the Milling Machine."
          The Smithsonian Journal of History, 1, No. 2
          (Summer 1966): 9-34.

     Rejects claims that Whitney invented the milling
machine, one of the fundamental components in the development of mass production methods. Credits Robert
Johnson with the invention. Concludes that Whitney
never perfected a system of interchangeable parts.

770. Green, Constance McLaughlin. *Eli Whitney and the Birth*
          of American Technology. Boston: Little, Brown,
          1956. viii + 215 pp. Bibliography, Index.

     Portrays Whitney as the inventor of the mass
production technique and hence the father of the industrial
North, as well as a major contributor to the development
of the Southern economy through the invention of the
cotton gin. This vision of Whitney has been sharply
attacked (see items 769 and 778), and Green's description
of Whitney as the inventor of interchangeable parts is
now viewed as the perpetuation of a myth.

771. Habakkuk, H.J. *American and British Technology in the*
          Nineteenth Century: The Search for Labour-Saving
          Inventions. 2nd edition. Cambridge: Cambridge
          University Press, 1967. ix + 222 pp. Index.

     Argues that the high cost and inelastic supply
of labor in the United States necessitated high productivity per unit of labor, leading to a relatively rapid
adoption of mechanization, standardization, and mass

production. Contends that Americans were not more
innovative than the British but were more focused in
their inventive efforts.

772. Saul, S.B. "Introduction." *Technological Change: The
           United States and Britain in the Nineteenth
           Century.*

        Evaluates Habakkuk's explanation of America's
rapid adoption of labor-saving devices (see item 771),
concluding that it is only partly correct. Calls for
industry-by-industry studies to identify patterns of
adoption of mass production techniques.

773. Sawyer, John E. "The Social Basis of the American System
           of Manufacturing." *Journal of Economic History,*
           14 (1954): 361-379.

        Calls for more emphasis on social explanations
in analyzing differential economic growth. Advocates a
central role for such explanations in the specific case
of the American System of Manufacturing.

774. Sinclair, Bruce. "At the Turn of the Screw: William
           Sellers, the Franklin Institute, and a
           Standard American Thread." *Technology and
           Culture,* 10 (1969): 20-34.

        Shows how the reputation of the Franklin
Institute for informed judgement and its links with the
nation's leading machinery firms enabled it to facilitate
the standardization of screws, an important step in the
evolution of mass production.

775. Smith, Merritt Roe. *Harpers Ferry Armory and the New
           Technology: The Challenge of Change.* Ithaca
           and London: Cornell University Press, 1977.
           363 pp. Bibliography, Index.

        Focuses on the resistance of the skilled
workers at the armory to the introduction of mechanization in the manufacturing of small arms, concluding that
the workers feared a reduction in status to mere machine
tenders, lower wages, and, generally, the imposition of
any concept identified as foreign to the local culture.
Describes the contributions of John H. Hall to the
development of the American System of Manufacturing while
at Harpers Ferry through the construction of machine
tools and quality control gauges.

776. ————. "John H. Hall, Simeon North, and the Milling
Machine: The Nature of Innovation Among Antebellum Arms Makers." *Technology and Culture*,
14 (1973): 573-591.

Demonstrates the role of the itinerant mechanic
in the transmission and dissemination of the mechanical
ideas essential for the transformation of the firearms
industry from craft to machine production.

777. Uselding, Paul. "Elisha K. Root, Forging, and the
'American System'." *Technology and Culture*,
15 (1974): 543-568.

Shows how a technique from one industry--ax-
making--was transferred to another--revolver manufacturing
--and in its new context became a key element in the
American System of Manufacturing.

778. Woodbury, Robert S. "The Legend of Eli Whitney and Interchangeable Parts." *Technology and Culture*,
1 (1960): 235-253.

Reduces claims that Eli Whitney was responsible
for the birth of the American System of Manufacturing
to myth.

ELECTRICITY AND ELECTRONICS

779. Belfield, Robert. "The Niagra System: The Evolution of
an Electric Power Complex at Niagra Falls,
1883-1896." *Proceedings of the IEEE*, 64
(1976): 1344-1350.

Describes the system as a synthesis of advanced
European and American technology, sufficiently flexible
to meet every type of consumer demand. Acknowledges
the aesthetic considerations which raised costs.

780. Bruce, Robert V. *Bell: Alexander Graham Bell and the
Conquest of Solitude*. Boston and Toronto:
Little, Brown and Company, 1973. xi + 564 pp.
Bibliography, Index.

Provides a highly detailed account of the life
of the inventor of the telephone. Argues that his
knowledge of accoustics and physiology was essential for
his success, as was the personal regard which his
financial partners had for him.

781. Conot, Robert. *A Streak of Luck*. New York: Seaview,
       1979. xvii + 565 pp. Bibliography, Index.

       Exploits archival material to reexamine the
life of Thomas Edison, in many cases substituting
documented facts for long-held myths. Pictures Edison
as a rather repulsive human being. The quality of the
analysis of Edison's personality and personal life is
superior to the technical discussions of his inventions.

782. Fink, Donald G. "Perspectives on Television: The Role
       Played by the Two NTSC's in Preparing Television
       Service for the American Public." *Proceedings
       of the IEEE*, 64 (1976): 1322-1331.

       Discusses the establishment of standards for
black-and-white (1940-1941) and color television (1950-
1953) by the National Television System Committees.

783. Hounshell, David A. "Bell and Gray: Contrasts in Style,
       Politics, and Etiquette." *Proceedings of the
       IEEE*, 64 (1976): 1305-1314.

       Argues that Alexander Graham Bell was victorious
in his priority dispute with Elisha Gray over the invention of the telephone because Bell had carefully presented
the telephone in the form of a scientific discovery,
obtained endorsements from the elder statesmen of science,
and utilized scientific institutions.

784. ———. "Elisha Gray and the Telephone: On the Dis-
       advantage of Being an Expert." *Technology and
       Culture*, 16 (1975): 133-161.

       Contends that Gray, a professional inventor
and expert on telegraphy, viewed the technology fundamental to the telephone as vital for multiple telegraphy
transmission rather than voice transmission. Argues
that Gray failed to patent the telephone because he saw
no immediate commercial use.

785. Hughes, Thomas P. "The Electrification of America: The
       System Builders." *Technology and Culture*, 20
       (1979): 124-161.

       Analyzes the contributions of three men, each
representative of a different stage in the development
of the electric power industry: Thomas Edison invented
electrical systems, Samuel Insull managed the resulting
light and power industries, while S.Z. Mitchell developed

the technique of financing regional systems through
holding companies.

786. Mabee, Carleton. *The American Leonardo: A Life of Samuel
F.B. Morse.* New York: Alfred A. Knopf, 1943.
xxi + 420 + xv pp. Index.

Presents a nonpartisan account of the life of
a man who had careers in art and technology. Sees Morse
as a man whose ignorance of the scientific world led him
to claim more credit than was actually due.

787. Norberg, Arthur L. "The Origins of the Electronics
Industry on the Pacific Coast." *Proceedings
of the IEEE*, 64 (1976): 1314-1322.

Finds a four-stage growth pattern: economic
colonialism through about 1910; a stable industrial
climate dominated by small, local entrepreneurs serving
markets too small for the Eastern firms; expanding
innovation resulting in a two-way flow of technical
information; technical maturity, accomplished around
1940. Discusses the first three stages in detail.

788. Passer, Harold Clarence. *The Electrical Manufacturers,
1875-1900: A Study in Competition, Entre-
preneurship, Technical Change, and Economic
Growth.* Cambridge, Mass.: Harvard University
Press, 1953. Reprint. New York: Arno Press,
1972. xviii + 412 pp. Bibliography, Index.

Studies the efforts of the engineer-entrepreneurs,
who combined the technical skill necessary to develop
marketable products with the business acumen needed to
recognize commercial possibilities. Considers only
arc lighting, incandescent lighting, and electric power.

789. Reynolds, Terry S., and Theodore Bernstein. "The
Damnable Alternating Current." *Proceedings
of the IEEE*, 64 (1976): 1339-1343.

Argues that the year 1888--the year Edison first
publicly attacked alternating current as unsafe--was the
critical year in the AC-DC debate. Finds that with the
technical and economic issues being decided in AC's
favor, the DC advocates focused, ultimately in vain,
on the safety issue.

790. Thompson, Robert Luther. *Wiring a Continent: The
                History of the Telegraph Industry in the United
                States, 1832-1866.* Princeton, Princeton
                University Press, 1947. Reprint. New York:
                Arno Press, 1972. xviii + 544 pp. Appendix,
                Bibliography, Index.

        Traces the history of the industry up to the
    establishment of the Western Union monopoly. Contends
    that most technical problems were solved through trial
    and error. Remains the definitive history.

IRON AND STEEL INDUSTRY

791. Allen, Robert C. "The Peculiar Productivity History of
                American Blast Furnaces, 1840-1913." *The
                Journal of Economic History,* 37 (1977):
                605-633.

        Credits the siliceous nature of American east
    coast ore with causing the mid-nineteenth-century gap
    in the productivity of American blast furnaces compared
    to those operated in Europe. Finds that the gap was
    rapidly closed in the 1870s and 1880s as a result of
    the switch to ore mined in the Lake Superior district
    which was less siliceous.

792. McHugh, Jeanne. *Alexander Holley and the Makers of Steel.*
                Baltimore and London: The Johns Hopkins Univer-
                sity Press, 1980. xiv + 402 pp. Appendices,
                Index.

        Provides both a biography of the man responsible
    for designing most of the Bessemer plants built in
    America during his lifetime and an overview of the
    history of steel making in the nineteenth century.
    Emphasizes the role of patents in influencing industrial
    decisions regarding competing technologies.

793. Schallenberg, Richard H. "Evolution, Adaptation and
                Survival: The Very Slow Death of the American
                Charcoal Iron Industry." *Annals of Science,*
                32 (1975): 341-358.

        Claims that the charcoal iron industry was able
    to survive until 1945 not, as usually argued, due to the
    abundance of wood in the United States, but because the
    industry was innovative and adaptive, fulfilling needs
    which the coal-smelt iron industry could not.

794. ——————, and David A. Ault. "Raw Materials Supply and
         Technological Change in the American Charcoal
         Industry." *Technology and Culture*, 18 (1977):
         436-466.

    Argues that the limiting factor in the improve-
ment and expansion of the industry was the quantity of
ore, not fuel. Critiques item 795, rejecting its
procedures.

795. Temin, Peter. *Iron and Steel in Nineteenth-Century
         America: An Economic Enquiry.* Cambridge,
         Mass.: The M.I.T. Press, 1964. ix + 304 pp.
         Appendices, Bibliography, Index.

    Attempts to explain the quantity of goods
produced and the price received by iron and steel
manufacturers through an analysis of changes in technical
requirements of production, the organization of production,
and the character of the demand.

TRANSPORTATION

796. Condit, Carl W. *The Railroad and the City: A Technological
         and Urbanistic History of Cincinnati.* Columbus:
         Ohio State University Press, 1977. xii +
         335 pp. Appendices, Bibliography, Index.

    Argues that in the wake of the railroad,
technology became the chief determinant of urban form,
with the rail network playing a major role in patterns
of land usage.

797. ——————. "Railroad Electrification in the United States."
         Proceedings of the IEEE, 64 (1976): 1350-1360.

    Divides the process into three stages: experimen-
tal; pioneering, marked by rapid progress in design,
construction, and operation (1895-1905); maturity.
Describes the process as piecemeal and fragmentary.

798. Emme, Eugene M., editor. *Two Hundred Years of Flight in
         America: A Bicentennial Survey.* AAS History
         Series, Volume I. San Diego: American Astro-
         nautical Society, 1977. xvi + 310 pp. Appendix,
         Index.

    Presents eight articles on ballooning, aviation
(general, military, and commercial), and spaceflight
(manned and unmanned). Most of the authors concentrate
on the technological development of the aircraft.

799. Flink, James J. *The Car Culture*. Cambridge, Mass. and
          London: The M.I.T. Press, 1975. x + 260 pp.
          Index.

          Proposes a three-stage history of the automobile
     in America. Sees the first stage, which marked the intro-
     duction of the automobile in this country, as terminating
     with the establishment of the Ford Highland Park plant in
     1910. Views the middle stage, 1910-1950, as the era of
     the mass idolization of the automobile; this was the
     period when the car culture was the foundation of
     America's consumer goods oriented economy. Argues that
     the final stage is characterized by the attitude that
     the automobile has become a social problem. Provides a
     view of the car in sharp contrast to that presented in
     item 804.

800. Hunter, Lewis C. *Steamboats on the Western Rivers: An
          Economic and Technological History*. Cambridge,
          Mass.: Harvard University Press, 1949. xv +
          684 pp. Appendix, Index.

          Presents an integrated treatment of the
     mechanical evolution, fiscal operation, and social and
     economic impact of the technology which greatly sped the
     development of the Western frontier. Demonstrates how
     the physical characteristics of the rivers determined
     many of the parameters of steamboat construction and
     operation. Argues that the mechanical development was
     the result of plodding progress rather than leaps of
     genius. Concludes that the steamboat reached its apex
     technologically during the post Civil War period, but
     that those safer, more comfortable, larger, and more
     efficient boats could not compete with the railroad
     system.

801. ———. "The Invention of the Western Steamboat." *The
          Journal of Economic History*, 3 (1943): 201-220.

          Rejects the single heroic inventor thesis in
     favor of viewing the Western steamboat as the result of
     evolutionary development through the efforts of numerous
     contributors.

802. Mazlish, Bruce, editor. *The Railroad and the Space
          Program: An Exploration in Historical Analogy*.
          Cambridge, Mass. and London: The M.I.T. Press,
          1965. ix + 233 pp. Index.

Includes seven essays which examine railroad technology, the impact of the railroad on the economy, the political system, and society, and the railroad as a cultural symbol. Attempts to forecast the impact of the space program upon American society through the use of historical analogy.

803. Rae, John B. *Climb to Greatness: The American Aircraft Industry, 1920-1960*. Cambridge, Mass. and London: The M.I.T. Press, 1968. xiii + 280 pp. Bibliography, Index.

Analyzes the industry prior to its transformation into the aerospace industry. Argues that military procurement policies led to technological competition among manufacturers and encouraged innovation.

804. ————. *The Road and the Car in American Life*. Cambridge, Mass. and London: The M.I.T. Press, 1971. xiv + 390 pp. Bibliography, Index.

Explores the economic and social impact of automotive highway transportation. Views the car and the highway as an integrated technological system. Reflects the highly optimistic vision that was widespread before high gasoline prices and shortages raised the question of conservation of resources.

805. Taylor, George Rogers. *The Transportation Revolution, 1815-1860*. New York and Toronto: Rinehart & Company, 1957. xvii + 490 pp. Bibliography, Appendices, Index.

Demonstrates how the new and improved transportation systems affected the economy of the United States. Discusses the impact of specific technological innovations. Includes an excellent, although now dated, critical bibliographic essay.

806. White, John H., Jr. *American Locomotives: An Engineering History, 1830-1880*. Baltimore: Johns Hopkins Press, 1968. xxiii + 504 pp. Appendices, Bibliography, Index.

Contends that locomotive design was governed by economics and the level of traffic. The low capital investment (resulting in inexpensive construction) and light traffic of American railroads demanded locomotives which were powerful, flexible, simple, inexpensive to construct, and easy to maintain; fuel efficiency and

speed were unimportant. Argues that standardized designs
were the rule by the middle of the century. Finds that
designers generally rejected innovation until the post-
1880 need for larger locomotives. Discusses the components
of locomotives, including boilers, gears, and cow catchers.
Provides many illustrations of representative locomotives.

807. ———. *The American Railroad Passenger Car*. Baltimore
and London: The Johns Hopkins University Press,
1978. xiii + 699 pp. Appendices, Bibliography,
Index.

Examines the design, construction, and use of
the standard railroad cars, with emphasis on those in
use during the nineteenth century. Discusses day coaches,
dining, parlor, sleeping, mail, baggage, and self-propelled
cars. Includes biographical sketches of designers and
builders, statistical data, and hundreds of illustrations.

MISCELLANEOUS INDUSTRIES

808. Anderson, Oscar Edward, Jr. *Refrigeration in America:
A History of a New Technology and its Impact*.
Princeton: Princeton University Press, 1953.
ix + 344 pp. Bibliography, Index.

Observes that the United States was both the
principal supplier of refrigeration technology and the
only major consumer who adopted it for general domestic
purposes. Argues that the felt need for food refrigeration arose about 1830 in response to changing American
diets and patterns of food supply; in turn, the advances
in technology enabled these dietary changes to become
more widespread. Focuses on the technologies for
harvesting and storing ice, refrigerated transportation,
and mechanical refrigeration. Demonstrates the impact
of technology on specific food industries, such as
meat packing.

809. Condit, Carl W. *American Building: Material and Techniques from the First Colonial Settlements to
the Present*. Chicago: University of Chicago
Press, 1968. xiv + 329 pp. Chronology,
Bibliography, Index.

Provides a comprehensive account of the effect
of building materials--wood, masonry, and metal--upon the
design of American buildings and the scientific advances
in our understanding of these materials. Focuses on

commerical and public buildings, bridges, dams, and train sheds. Discusses the limitations of the construction material, the needs each could fill, and the interaction of material and aesthetic considerations.

810. Enos, John Lawrence. *Petroleum Progress and Profits: A History of Process Innovation*. Cambridge, Mass.: The M.I.T. Press, 1962. xiii + 336 pp. Appendices, Bibliography, Index.

Discusses six specific cracking-process innovations, focusing on the need for new technology, the research and development process, the advantages of the new technology, and the effect on the industry.

811. Feller, Irwin. "The Diffusion and Location of Technological Change in the American Cotton-Textile Industry, 1890-1970." *Technology and Culture*, 15 (1974): 569-593.

Tests and rejects the thesis that a developing economy is superior to a mature one in the utilization of technological innovations. Finds that the exploitation of innovations was about equal in the older cotton-textile mills of New England and the newer mills of the South. Concludes that it was the lower salaries paid in the South, not a reluctance by New England mill owners to update equipment, which enabled Southern mills to be more competitive.

812. Haynes, Williams. *American Chemical Industry*. Volume I: *Background and Beginnings, 1609-1911*. Volume II and Volume III: *The World War I Period, 1912-1922*. Volume IV: *The Merger Era, 1923-1929*. Volume V: *The Decade of New Products, 1930-1939*. Volume VI: *The Chemical Companies*. New York, Toronto, and London: D. Van Nostrand Company, 1945-1954. lxxvii + 512 pp., xliii + 440 pp., xv + 606 pp., xli + 638 pp., li + 622 pp., viii + 559 pp. Appendices, Bibliographies, Indices, Errata.

Provides a detailed overview of the American chemical industry prior to World War II. Discusses both the technical developments and the corporate maneuvers. Remains a very useful, although somewhat celebratory reference work.

813. Jenkins, Reese V. *Images and Enterprise: Technology and*
     the American Photographic Industry, 1839 to
     1925. Baltimore and London: The Johns Hopkins
     University Press, 1975. xviii + 370 pp.
     Appendix, Bibliography, Index.

     Demonstrates the relationship between technolog-
ical innovation and changes in business organization,
while tracing the evolution of the American photographic
industry from a small diffuse industry to a large,
centralized one with heightened awareness of the role
of technology in business strategy. Identifies five
major technological stages in photography: daguerreotypes,
wet collodions, dry gelatin on glass plates, gelatin on
celluloid rolls, and cinematographic film. Argues that
each stage was marked by a three-phase market structure:
imperfect competition (a few innovators), perfect competition (many imitators enter the field), and oligo-
polistic competition (domination by a limited number
of firms).

814. ———. "Technology and the Market: George Eastman and
     the Origins of Mass Amateur Photography."
     Technology and Culture, 16 (1975): 1-19.

     Argues that the transformation of photography
during the last quarter of the nineteenth century from a
professional to an amateur field was the result of the
combination of a series of interrelated changes in
photographic technology, especially the development of
the roll film system, with certain marketing strategies.

815. Jeremy, David J. "British Textile Technology Transmission
     to the United States: The Philadelphia Region
     Experience, 1770-1820." *Business History Review*,
     47 (1973): 24-52.

     Argues that the lack of printed sources for
textile machinery construction and operation placed the
burden of the technology transfer upon the memory and
skills of immigrant artisans.

816. ———. "Innovation in American Textile Technology During
     the Early 19th Century." *Technology and Cul-*
     ture, 14 (1973): 40-76.

     Demonstrates how the combination of a search for
a market relatively free from competition from British
imports or household manufacturers and the necessity of
utilizing women to meet the shortage in unskilled labor

led the textile industry to embrace particular labor-
saving principles. Examines the specific innovations
created to incorporate these principles.

817. Roberts, William I., III. "American Potash Manufacture
Before the American Revolution." *Proceedings
of the American Philosophical Society*, 116
(1972): 383-395.

Blames the underdevelopment in the colonies of
the principal eighteenth-century chemical industry on a
lack of technical know-how, not unfavorable economic
conditions.

818. Wagoner, Harless D. *The U.S. Machine Tool Industry from
1900 to 1950*. Cambridge, Mass.: The M.I.T.
Press, 1968. xiv + 421 pp. Bibliography,
Indices.

Provides an overview--qualitative and quantita-
tive--of the development of the industry. Argues that the
critical problems facing the industry during these years
were managerial--adjusting production, sales, and
administration to changing economic and political con-
ditions--not technical.

819. White, Gerald T. *Scientists in Conflict: The Beginnings
of the Oil Industry in California*. San Marino:
The Huntington Library, 1968. xiii + 272 pp.
Bibliography, Index.

Describes the California oil boom of 1865 and
the resulting controversy between Benjamin Silliman, Jr.,
the scientist whose reports triggered the boom, and
Josiah D. Whitney, the director of the California
Geological Survey, who had concluded that the oil deposits
in California were not commercially valuable. Defends
Silliman against charges of salting the oil samples.
Places the blame for the collapse of the California
Geological Survey on Whitney's generally inept public
and political relations, rather than the specific issue
of whether the Survey had committed a major scientific
error while analyzing the oil deposits. Discusses the
refusal of the scientific community to condemn Silliman
publicly.

820. Williamson, Harold, et al. *The American Petroleum
        Industry.* Volume I: *The Age of Illumination,
        1869-1899.* Volume II: *The Age of Energy, 1899-
        1959.* Evanston: Northwestern University Press,
        1959-1963. xvi + 864 pp., xx + 928 pp.
        Indices, Appendices.

    Provides a detailed, integrated history of the
industry. Discusses business organization and competition,
technological developments in drilling and refining,
changes in public demand, and the role of government.

SPACE EXPLORATION

821. Brooks, Courtney G., James M. Grimwood, and Loyd S.
        Swenson, Jr. *Chariots for Apollo: A History
        of Manned Lunar Spacecraft.* The NASA History
        Series. Washington, D.C.: National Aeronautics
        and Space Administration, 1979. xvii + 538 pp.
        Appendices, Bibliography, Index.

    Traces the evolution of the hardware that
carried man to the moon. Includes highly detailed
discussions of changes in design. Focuses on the debate
over the mode of travel to the moon (direct ascent,
rendezvous, or parking orbit). Discusses managerial
problems, especially the relationship between NASA and
its contractors. Offers little insight or information
on the role of science in the Apollo program.

822. Dickson, Katherine Murphy. *History of Aeronautics and
        Astronautics: A Preliminary Bibliography.*
        Washington, D.C.: National Aeronautics and
        Space Administration, 1968. viii + 420 pp.
        Indices.

    Provides descriptive annotations of both
primary and secondary printed material.

*   Emme, Eugene, M., editor. *Two Hundred Years of Flight
        in America: A Bicentennial Survey.* Cited
        above as item 798.

    Contains essays on manned space flight and
instrumental exploration of space by Edward C. Ezell
and R. Cargill Hall, respectively.

823. Green, Constance McLaughlin, and Milton Lomask.
     Vanguard: A History. The NASA Historical
     Series. Washington, D.C.: National Aeronautics
     and Space Administration, 1970. xvi + 308 pp.
     Bibliography, Appendices, Index.

     Describes the conception, development, and
results of the first American satellite program. Emphasizes the difficulties produced by interservice rivalry.
Suggests that the project's low priority relative to the
military ballistic missile program was another source of
problems. Concludes that the program was ultimately a
success from both the scientific and engineering
perspectives.

824. Hall, R. Cargill. *Lunar Impact: A History of Project
     Ranger*. The NASA History Series. Washington,
     D.C.: National Aeronautic and Space Administra-
     tion, 1977. xvii + 450 pp. Appendices, Index.

     Analyzes the causes and impact of the struggle
between engineers and scientists and between sky and
planetary scientists over the establishment of priorities
within the NASA program. Assesses the differences
between academic and industrial laboratory managerial
styles as it bore upon the conflict between the Jet
Propulsion Laboratory and NASA over the administration
of Ranger. Concludes that Ranger taught NASA the
importance of centralized control and the dependence
of successful space exploration upon the ascendancy
of engineering over scientific priorities.

825. Hallion, Richard P., and Tom D. Crouch, editors. *Apollo:
     Ten Years Since Tranquility Base*. Washington,
     D.C.: National Air and Space Museum, 1979.
     xviii + 174 pp. Bibliography.

     Presents eighteen essays dealing with the
political, economic, social, scientific, and technological
aspects of the Apollo program. Most pieces are too short
to do justice to their subject matter.

826. Hartman, Edwin P. *Adventures in Research: A History of
     Ames Research Center, 1940-1965*. NASA Center
     History Series. Washington, D.C.: National
     Aeronautics and Space Administration, 1970.
     xviii + 555 pp. Appendices, Index.

     Traces the physical evolution, describes the
staff, and examines the achievements of the center, a

major site for aerospace research. This work has only
minimal documentation and was written by an aerospace
engineer with admitted biases towards the men and
events under discussion.

827. Lay, Beirne, Jr. *Earthbound Astronauts: The Builders of
Apollo-Saturn.* Englewood Cliffs: Prentice-
Hall, 1971. vii + 198 pp.

Argues that the keystone of the Apollo-Saturn
Project was the engineer. Based on interviews, this is
an undocumented, biased work, which borders on an
exercise in hero-worship.

828. Lehman, Milton. *This High Man: Life of Robert Goddard.*
New York: Farrar, Straus, 1963. xv + 430 pp.
Bibliography, Index.

Utilizes manuscripts and interviews to detail
the life and research of the American pioneer in liquid-
fuel rocketry. Presents a sympathetic view of the man;
this is the official biography. Discusses the mixture of
foundation (during the interwar years) and military
support which Goddard used to fund his work. Views
Goddard as essentially a solitary researcher, uncomfortable
in the limelight.

829. Lewis, Richard S. *Appointment on the Moon.* Revised
Edition. New York: Ballantine Books, 1969.
568 pp. Index.

Offers a journalistic survey of the American
space program, emphasizing the civilian manned projects.
Concentrates on the human element rather than the
hardware or scientific research.

830. Logsdon, John M. *The Decision to Go to the Moon: Project
Apollo and the National Interest.* Chicago and
London: The University of Chicago Press, 1976.
xiii + 187 pp. Index.

Views the decision by John F. Kennedy as a
political judgment to use technology for national purposes.
Describes the aim as the achievement of a spectacular
success in order to enhance the national prestige. Argues
that a lunar landing was chosen because it was technically
feasible, yet beyond the capabilities of the contemporary
generation of both American and Russian rockets, negating
the momentary Russian advantage in booster power. Concludes that the "Apollo approach" could be used to attack

other national problems if such problems were susceptible
to a technological fix, the proper crisis atmosphere
existed, and the political leadership was present.

831. Newell, Homer E. *Beyond the Atmosphere: Early Years of
Space Science*. The NASA History Series.
Washington, D.C.: National Aeronautics and
Space Administration, 1980. xviii + 497 pp.
Appendices, Bibliography, Index.

Examines the conduct of scientific investigations
assisted by rockets, satellites, and space probes from
the perspective of NASA headquarters. Concentrates on
internal administration, the interaction of NASA and
other executive agencies, and the "love-hate" relationship
between NASA and academic scientists. Provides detailed
discussions of the contributions of space science to
studies of the magnetosphere and geodesy.

832. Swenson, Loyd S., Jr. "The 'Megamachine' Behind the
Mercury Spacecraft." *American Quarterly*, 21
(1969): 210-227.

Uses Lewis Mumford's conception of the
"megamachine"--a massive social organization operating
as an engine of large-scale construction or destruction--
to investigate the structure of the government-industry
team which developed the technology for the Mercury
Project.

AUTHOR INDEX